Charles Otis Whitman

Methods of Research in Microscopical Anatomy and Embryology

Charles Otis Whitman

Methods of Research in Microscopical Anatomy and Embryology

ISBN/EAN: 9783744690201

Printed in Europe, USA, Canada, Australia, Japan

Cover: Foto ©berggeist007 / pixelio.de

More available books at **www.hansebooks.com**

METHODS OF RESEARCH

IN

MICROSCOPICAL ANATOMY

AND

EMBRYOLOGY

BY

CHARLES OTIS WHITMAN, M.A., Ph.D.

ILLUSTRATED

Boston
S. E. CASSINO AND COMPANY
1885

PREFACE.

The aim of this work is to supply, in a measure, a need which has been created by the rapid development of the methods of microscopical anatomy and embryology within the last few years. No effort has been made to give it an encyclopædic character, or to render it an exhaustive treatise of the subject in any of its aspects. With avalanches of methods constantly sweeping down upon us, it would be as futile as it is undesirable to attempt to enclose the whole field. We are already richly supplied with books on the microscope, which deal with its history, optical principles, forms of construction, accessories, modes of use, etc.; and histological descriptions are furnished in numerous systematic works. The important branch of methods represented by microphotography has also been handled in special works.

Pre-occupation of the ground in these directions is not, however, the only reason for the limitations imposed on this volume. Microscopical technique is no more closely connected with histological than embryological studies; and the practical considerations which would certainly make it advisable not to encumber an embryological treatise with a general survey of methods appear to me to hold equally in the case of histological text-books. As the embryologist and histologist employ, for the most part, identical methods and instruments, separate treatment of these is obviously an advantage which may be shared by both. Hitherto, most of our standard books of reference on methods have been rather complex in character, dealing with the microscope and technical methods as subordinate and introductory to the main subject of histology. Works of this description, however invaluable as histological manuals, are not well adapted to the every-day needs of the zoological laboratory. These needs may be summed up in two words, namely, *material* and *methods*.

Under the head of material may be included, not only the object of study, but also the times, places, and best means of collecting, as well as breeding habits, food, etc.; in short, all the items of information that would aid the student in making a *choice* of material and in controlling its *supply*. The requirements of the histologist in this direction have been bountifully provided for; while those of the embryologist have been almost wholly ignored. The embryologist is not, to be sure, wholly destitute of sources of information; but these sources are scattered and inaccessible, and the information they offer is, for the most part, too meagre to be of great assistance. Embryologists are themselves, in a large measure, responsible for the inadequate supply of this kind of information; for it is only quite recently that the matter has received any proper attention at their hands, and even now it is often treated with the greatest indifference. In the course of any complete and thorough study of the development of an animal, the investigator acquires a fund of information on some or all of the points above named, which is not, as a rule, to be obtained from systematic works, and which may be of the highest value to the observer who next takes up the same subject. Neglect to give such information is, at the best, inexcusable; and to withhold it on the ground that others are not entitled to the advantages of your experience is nothing less than contemptible.

The importance which I attach to this point is not to be measured by the space devoted to it in the following pages; for the time which I could give to it has been from necessity very limited, so that I have scarcely done more than indicate what should be done on a much larger scale. Should the plan commend itself to the judgment of others, it may be more fully developed hereafter.

This class of facts forms a part of the embryological methods, embracing some notes on the occurrence of eggs appended thereto.

The contents of this volume have been arranged in two parts, the first embracing methods of a more general nature, such as preservative fluids, dyes, macerating fluids, fixatives, mounting media, the microtome with its appurtenances, methods of imbedding, etc.; the second including special applications of embryological, anatomical, and histological methods. In an appendix will be found some methods of injection, museum methods, and formulæ for most of the important reagents, etc.

The introduction of special methods may perhaps be regarded as a not unimportant feature of the present volume. This part is designed to meet the wants of the beginner, as well as the more advanced student; and, under the persuasion that it will be found useful, I have only to regret my present inability to make it much more complete.

Sources of indebtedness have been acknowledged in the text; but it should be mentioned that the matter here presented has been drawn, to a large extent, from summaries and notes that have appeared in the *American Naturalist* during the past three years, and from a paper on the methods used at the Naples Zoological Station, for the substance of which I was so largely indebted to Paul Mayer. This material has, however, been entirely rearranged, and additions and alterations have been made in all directions, in order to bring it up to the latest improvements and discoveries.

In this work of revising and supplementing, I have been materially aided by the excellent and comprehensive summaries which appear in every volume of the Naples "Zoologischer Jahresbericht," from the pen of Max Flesch of Würzburg. In a similar manner, I have made frequent use of the "Journal of the Royal Microscopical Society," edited by Frank Crisp of London, and the new "Zeitschrift für wissenschaftliche Mikroskopie," published by Wilh. Jul. Behrens in Göttingen.

Fol's Lehrbuch der Vergleichenden Mikroskopischen Anatomie (erste Lieferung) has also been often consulted, as will appear from numerous references.

I am under obligation to Paul Mayer, Ryder, Kingsley, Minot, Mark, Faxon, Osborne, and Wilson for matter which appears for the first time in these pages, and to the *American Naturalist* for most of the cuts. To Mr. Kingsley I am especially indebted for correcting the proof-sheets, and for many suggestions that have proved very valuable.

C. O. WHITMAN.

MUSEUM OF COMPARATIVE ZOOLOGY,
Cambridge, 1885.

CONTENTS.

PART I.—GENERAL METHODS.

	PAGE
INTRODUCTION	1

CHAPTER I.—METHODS OF KILLING, HARDENING, PRESERVING, BLEACHING, MACERATING, DECALCIFYING, AND DESILICIFYING 11

  II.—METHODS OF STAINING 33

  III.—INSTRUMENTS 60

  IV.—METHODS OF IMBEDDING 93

  V.—FIXATIVES FOR SERIAL SECTIONS 116

  VI.—MOUNTING MEDIA 124

  VII.—THE USES OF COLLODION 126

PART II.—SPECIAL METHODS.

CHAPTER VIII.—EMBRYOLOGICAL METHODS 129

  IX.—TIMES AND PLACES OF OVULATION 174

CHAPTER X. — KARYOKINETIC FIGURES 181

XI. — PREPARATION OF NERVE-TISSUE 187

XII. — HISTOLOGICAL METHODS 200

XIII. — RECONSTRUCTION FROM SECTIONS 213

APPENDIX.

1. METHODS OF INJECTION 219

2. MUSEUM METHODS 231

3. FORMULÆ FOR REAGENTS, DYES, ETC. 237

INDEX 249

Part I.

GENERAL METHODS.

MICROSCOPICAL METHODS.

INTRODUCTION.

In beginning the study of any branch of natural history, the student of to-day finds that the first step in preparation for independent investigation is to become familiar, through actual laboratory work and under the guidance of experienced instructors, with the more important instruments and methods of microscopical research. This is especially true in every department of zoology. Even in systematic zoology, it cannot be neglected with impunity. Macroscopical and microscopical methods of study are independent of each other only to a very limited degree; they have a hand-in-hand relation in every thorough and complete morphological study, sometimes parting company it is true, but only to supplement and strengthen each other in the end. The controversies on the question of using chemical reagents in the investigation of animal or vegetable tissue have ceased to have more than a historical interest. Karyokinetic figures are no longer regarded with suspicion, and the "Trugbilder" of metallic impregnations are now universally approved as reliable pictures of conditions that actually exist in living tissues. Methods of study are now cultivated with as much zeal by the embryologist as the histologist, and the anatomist and the physiologist feels compelled to supplement their methods by those of the 'plodding microtomist.' Microscopical methods are not then the monopoly of any particular class of scientific investigators.

The methods of microscopical research now employed are, to a very large extent, the product of the experience of the ablest investigators in various departments of biology, during the last ten or fifteen years. So rapid has been the progress in this direction, that it has been found necessary to supplement our hand-books of methods through the publication of journals devoted exclusively to the work of recording and summarizing new developments in zoological technique.

The rapid development of methods is at once the result and one of the chief causes of the increasing activity in every field of biological research. The improvement of methods leads to the re-investigation of old subjects, and at the same time prepares the way for attacking new problems. The investigator who neglects to keep himself informed of the progress in methods of study throws away his opportunities, and has the vexatious mortification of seeing himself outdone and his work superseded by that of more skilful hands.

The microtome has come to occupy a place in the zoological laboratory second in importance only to the microscope itself. Many improvements in details and in accessories have followed the introduction of this instrument, and a whole series of methods has sprung up in connection with its use. In short we have a new art, which has been appropriately called *microtomy*.

The general favor with which the microtome has been received is the best evidence of its usefulness. The use of the instrument is so simple, and the methods connected with it so easily acquired, that no naturalist can afford to work without it.

Within the last four or five years the improvements and discoveries in microtomy have been both numerous and extremely important. Among these may be mentioned the ribbon method of cutting serial sections, discovered by Caldwell; the methods of fixing sections on the object-slide, discovered by Giesbrecht, Schällibaum and Mayer; the various section-smoothers, notably those invented by Mayer, Andres, Giesbrecht, Schulze and Decker; the use of collodion to prevent the crumbling of brittle sections, proposed by Mason; and the methods of reconstructing objects from serial sections employed by His and Born. A large number of new preservative and staining fluids have been described; and new methods of killing, hardening, preserving, staining, and imbedding have been recommended.

Experience has shown that each different object requires a special mode of treatment, and that the same object must be treated differently according to the nature of the problem in hand. For example, the course of preparation which has given satisfactory results in the study of the development of the ova of a certain species may prove quite inadequate when applied to a different though closely allied form; and it has been found that different stages in the development of the same ovum often require different modes of preservation. The investigator cannot, therefore, blindly adopt the methods employed by others, but must, in by far the greater number of cases, determine by experiment the method to be pursued. But such experiments demand a general knowledge of methods, and, above all, a knowledge of the special applications of methods in cognate subjects. It is in the adaptation of methods to special subjects that the skill of the investigator is shown. Our information of the methods employed in specific cases should therefore be as extended as possible. To meet this need, entire courses of methods that have led to successful results in typical cases have received attention in the second part of this volume.

A few explanatory remarks and general directions may enable the beginner to enter upon an experimental study of methods with his attention at least properly oriented.

From a general standpoint, we may say that there are two modes of studying an animal, an organ, a tissue, or a cell, which we may distinguish as *surface-observation* and *section-observation*. In surface-observation, the object may be examined in a living (or fresh) condition, or after treatment with preservative fluids; in section-observation — leaving the freezing methods aside — reagents must be employed, and the various steps in the preparation must of course follow some definite order or sequence. After learning this sequence in a single case and mastering the principles underlying it, the beginner will be prepared to proceed intelligently with other cases. An illustration will put this matter in a clearer light.

A small worm is a suitable and convenient object of study by means of serial sections. After making the preliminary surface examination, with a view to ascertaining the external features and their precise topographical relations, the preparation for sectioning may be undertaken. In the usual order of procedure, the several steps are, (1) *killing*, (2) *hardening*, (3) *staining*, (4) *dehydration*, (5) *clarification*, (6) *saturation with paraffine*, (7) *imbedding*, (8) *sectioning*, (9) *arrangement and fixation on the slide*, and (10) *inclusion in balsam*.

The first two steps are often accomplished by a single reagent; and sometimes the third is combined with the first two, either by using a reagent (*e. g.* osmic acid) that will do three kinds of work at the same time, or by mixing the staining fluid with the killing and hardening medium (Perenyi's fluid). The third step may also be suppressed, the sequence otherwise remaining the same; or it may be transposed so as to follow the ninth, in which case the fourth and fifth would have to be repeated before concluding with the tenth. Excepting the simple modifications by *combination, suppression,* and *transposition*, here mentioned, the above sequence may be said to be invariable. The real field for experimentation then lies, not in the order of the steps, but in the mode of executing them: and, beyond the first three steps, the method of procedure is so free from complications that the beginner will need very little assistance in following the course prescribed.

With so much of a general nature premised, we may take up the consideration of the first three steps of the preparation in their usual order, noticing, as we pass, a few of the more important ways of varying the course, and calling attention to such points as will tend to lessen the chances of aimless experimentation.

1. Killing. The all-important point here is to preserve the natural form in an extended condition, with as little alteration of the histological elements as possible. This can usually be accomplished in one of two general ways: — first, by immersion in a fluid that kills instantly; second, by using one that operates so slowly that sensibility is lost very gradually. In some cases either method may be employed with about equal success, while in others the advantages may be largely or wholly on one side. In order to make an intelligent choice among the various agents employed for killing, it is necessary to know something of their peculiarities, and then to select with reference to the character and size of the animal, and the special purpose in view.

If the worm be a Planarian, the method of killing instantaneously with corrosive sublimate would probably give the most satisfactory results. The same method could be employed with most annelids; but here our preference might take a somewhat wider range, extending to one or more of the following agents,— alcohol added slowly as recommended by Eisig, Merkel's chrom-platinum solution, chromic acid, or Kleinenberg's picro-sulphuric acid. These five methods are by no means all that might be resorted to; but the cases in which the desired results could not be obtained through one or more of them, are too exceptional to call for notice here.

Corrosive sublimate kills and hardens at the same time, and accomplishes its work in a few minutes. It kills in most cases so quickly that the animal has no time to contract; and it leaves the tissues receptive to staining fluids. The moment the solution has thoroughly penetrated all parts (10–30 m.), its action should be interrupted by washing in distilled water until the sublimate has been entirely removed (30–60 m.). If the washing is not thorough, needle-like crystals will form and impair or ruin the preparation.

Alcohol added by drops is a somewhat slower process, requiring very close watching. The chief advantage of this method is that it leaves no obstacles in the way of staining. The dangers to be carefully guarded against are, (1) *contraction*, in consequence of a too rapid addition of alcohol, and (2) *maceration*, resulting from an undue prolongation of the process. As soon as the animal is dead, it should be placed for a short time (30 m.) in 40 % alcohol, then in 50 %, 60 %, 70 %, 80 %, 90 %, doubling the time at each step.

Picro-sulphuric acid has the admirable quality of causing little or no shrinkage, and of being easily removed in moderately strong alcohol; but in most cases direct immersion in this fluid would kill the worm in a contracted condition. The danger in this direction can only be met by a preliminary treatment with some agent that temporarily or permanently deprives the worm of the power of contraction. Chloroform would serve this purpose in many, though not all cases.

Chromic acid is a convenient and much used reagent for killing and hardening. If a worm be placed in a weak solution, it may show vigorous contractions for a considerable time, but in the end it generally dies in an extended condition. If it requires to be straightened, this should be done immediately, before the hardening process has rendered it inflexible.

The great drawback in the use of chromic acid is the difficulty of staining after it. This obstacle cannot be wholly overcome, but it can be lessened considerably by interrupting the action of the acid soon after the animal is completely dead. This reagent hardens by virtue of chemical combinations which it forms with cell substances, and it is this peculiar hardening action which renders the tissues non-receptive to staining fluids. The sooner, therefore, this action is interrupted by washing, the easier will be the work of staining. It is advisable to use a weak solution (not over $\frac{1}{8}$ %) and to limit its action so far as possible to killing, leaving the hardening to be accomplished by alcohol of 40 %, 50 %, 70 %, and 80-90 %.

Merkel's fluid is chromic acid tempered somewhat in its action by platinum chloride. As a killing agent it is superior to chromic acid, but as a hardening agent it has the same objectional feature. The suggestion made above in regard to limiting the action of chromic acid, and leaving the work of hardening to alcohol, holds with equal force and for similar reasons in the case of Merkel's fluid.

2. Hardening. The work of hardening, after each of the foregoing methods of killing, will fall mainly or wholly to alcohol. The chief danger here lies in shrinkage, to avoid which it is necessary to use different grades of alcohol, advancing gradually to 80 % or 90 %. It is generally safe to begin with 40 % or 50 %, and sometimes only two grades (70 % and 90 %) are required. The strength of the first grade, as well as the number of grades to be interposed between this and 90 %, will depend on the method of killing and the delicacy of the object.

The use of ascending grades of alcohol to avoid the shrinkage that would follow direct transference to 90 %, is a necessity which finds its explanation in the attractive power of alcohol for water. On exposure to alcohol the water of the tissues is withdrawn more rapidly than it is replaced by alcohol; and the stronger the alcohol the more unequal the diffusion. The aim, then, in beginning with weak alcohol and advancing gradually, is to tone down the violence of the osmotic action, and thus lessen the danger of collapse.

In the use of alcohols below 60 %, considerable care must be taken to guard against maceration: this danger must be avoided by transferring as rapidly as possible through these grades.

It is impossible to give precise and infallible directions for accomplishing this step, for no inflexible system of rules can be devised to meet all the possible emergencies. The following indications of *time* and *strength* are therefore to be regarded as mere approximations, having very little value beyond the aid they may give the experimenter in making his first trials.

Assuming that the object is not more than a few centimeters long, and from two to four millimeters thick, the corrosive sublimate preparation, after being washed for an hour in water supplied in abundance and four or five times renewed, may be transferred to alcohol of 50 % one hour, 60 % two hours, 70 % twelve to twenty-four hours, 80–90 % twenty-four hours.

The picro-sulphuric acid preparation, after remaining from one to two hours (till completely penetrated) in the acid, may be transferred directly, without washing, to 70 % alcohol for twenty-four hours, and the alcohol changed four or five times (until every trace of the yellow color is removed from the object); then to 90 % for twenty-four hours or longer.

The preparations in chromic acid and Merkel's fluid, after an immersion of one-half to one hour, should be thoroughly washed (best in running water) for about the same time, then placed in 40 % alcohol fifteen minutes, 50 % thirty minutes, 60 % one hour, 70 % five hours, and 80 % ten to twelve hours.

The treatment of the alcoholic preparation has already been indicated.

3. Staining. The work of staining will now be in order, if the object admits of being stained as a whole. The postponement of this step till the sections have been made and fixed on the slide may be considered as an alternative, to be resorted to in cases of necessity, *i. e.*, when staining *in toto* is found to be ineffectual. If the worm is three or four centimeters in length, it will facilitate the staining considerably to divide it into two or three parts; and such a division will be found to be an advantage in the further preparation, as well as a necessary preliminary to imbedding.

As to the kind of dyes to be employed, it may be said that *alcoholic solutions*, such as Grenacher's borax-carmine, Kleinenberg's hæmatoxylin, &c., are, as a rule, to be preferred to *aqueous solutions*, such as Beale's carmine, picro-carmine, &c. The grounds for this preference will be discussed under the head of staining fluids, and need not therefore occupy us here.

Among the more important qualifications of a dye for staining *in toto* are the following: —

1. It should *penetrate* readily, so that its action on the deeper and more superficial parts may be as equable as possible.

2. Its action, at full strength, should be *intense* — capable of imparting any desired depth of color.

3. It should give a *differential* rather than a *diffuse* staining; *i. e.*, it should act with unequal degrees of intensity on the different elements of the cell, thus contributing to *sharpness of definition*.

4. Its *preservative* qualities should be sufficiently pronounced to enable it to act any required length of time without causing maceration.

5. The color imparted should be *permanent*.

All these favorable qualities are combined in a very high degree in Kleinenberg's hæmatoxylin; and, to a somewhat less extent, in borax-carmine. In richness and *transparency* of tint the carmine solution has the advantage; but in point of penetrancy and freedom from danger of maceration, hæmatoxylin takes the higher rank. In using these dyes it must be remembered that the hardened objects should not be transferred directly from 90 % alcohol, but should first be left for half an hour or more in alcohol of the same strength as the coloring solution. It should also be borne in mind that every trace of acid must be removed before immersing the object in hæmatoxylin, for the color will inevitably fade if any free acid is left in the tissues.

The preparations in sublimate, picro-sulphuric acid, and alcohol, will require only a few hours for staining, and hence the choice of dyes, assuming that it is limited to these two, may be made according to preference for color. The preparations in chromic-acid and Merkel's fluid will stain with greater difficulty; but if the time required does not much exceed twenty-four hours, borax-carmine can be used with perfect safety, provided the hardening has been successfully accomplished. The hæmatoxylin solution will stain quicker than the carmine, and in many cases would undoubtedly give the more satisfactory results.

No precise time-limits can be prescribed for the action of the coloring fluid: the preparations must be examined from time to time, and removed from the dye when they appear to be thoroughly and deeply stained. They should be washed in alcohol of the same grade as the dye, and then dehydrated by treatment with 70 %, 90 %, and absolute alcohol.

The differential character of the stain may be strengthened by washing a few minutes (before dehydrating) in slightly acidulated alcohol.

It has already been mentioned that the difficulty of staining chromic acid preparations may be lessened by cutting short the action of the acid as soon as its penetration to all parts is assured. There is still another precaution, the observance of which will tend to reduce the evil to a minimum. *The staining should follow at the earliest possible moment after the action of the acid; for the difficulty of staining increases with the lapse of time.* The second step of hardening should therefore be shortened rather than prolonged, so that the first and third steps may not be separated by a longer interval than necessary.

The remaining steps of the preparation present no complications that require to be considered here. The directions for these in the following pages will, it is presumed, be found sufficiently full to enable the beginner to proceed independently.

Microtomic methods, it should be remembered, form only a part of the methods employed in the study of animal morphology. That they constitute a very important part is certain; but it is equally certain that surface-observation must not be neglected. Cultivation of either class of methods to the exclusion of the other accounts for the unsatisfactory results of many an investigation. Over-confidence in section-making *may* prove as disastrous as total ignorance of it. It would be easy to cite examples on both sides. Surface-observation and section-observation go hand in hand, and the highest advantages of either are known only to those who know how to practise both. The work of the microtome is supplementary in character; its function is to verify and correct, to strengthen the weak points and supply the deficiencies of surface-observation.

CHAPTER I.

PRESERVATIVE FLUIDS.

Killing, hardening, and preserving are three kinds of work, requiring for their accomplishment sometimes only a single preservative fluid, but in most cases two, three, or even more. As the same fluid often does the work of killing and hardening, and sometimes of preserving too, it is impossible to divide them into three classes corresponding to the kinds of work, except by repeating many of them twice, and some of them three times. While it is therefore more convenient to include them all under "preservative fluids," as Dr. Mayer[1] has done, it is none the less important to remember what kind of work each fluid is expected to accomplish.

Kleinenberg's picro-sulphuric acid, for instance, is not a hardening fluid. It serves for killing, and thus prepares for subsequent hardening.

HOT WATER.

Water heated to about 80° C., as first recommended by Bobretzky, is an excellent mode of killing some kinds of eggs, especially those of many Arthropoda. This method has also been employed to advantage with the amphibian egg.

ALCOHOL.

In the preparation of animals or parts of animals for museums or histological study, it is well known that the chief

---

[1] "Mittheilungen aus der Zoologischen Station zu Neapel," Vol. 2, p. 1, 1880.

difficulties are met in the process of killing. Alcohol, as commonly used for this purpose by collectors, has little more than its convenience to recommend it. P. Mayer has called attention to the following disadvantages attending its use in the case of marine animals: —

(1) In thick-walled animals, particularly those provided with chitinous envelopes, alcohol causes a more or less strong maceration of the internal parts, which often ends in putrefaction.

(2) In the case of smaller Crustacea, *e. g.*, Amphipods and Isopods, it gives rise to precipitates in the body fluids, and thus solders the organs together in such a manner as often to defy separation, even by experienced hands.

(3) It fixes most of the salts of the water adhering to the surface of marine animals, and thus a crust is formed which prevents the penetration of the fluid to the interior.[1]

(4) This crust also prevents the action of staining fluids, except aqueous solutions, by which it would be dissolved.

Notwithstanding these drawbacks, alcohol is still regarded at the Naples Station as an excellent fluid for *killing* many animals designed for preservation in museums or for histological work. In many cases the unsatisfactory results obtained are to be attributed, not to the alcohol *per se*, but to the *method* of using it. Most of the foregoing objections do not, as Dr. Mayer has expressly stated, apply to fresh-water animals; and Dr. Eisig informs me that he has no better method of killing marine annelids than with alcohol.

Eisig's Method of Killing with Alcohol. — The process is extremely simple. A few drops of alcohol are put into a vessel which contains the annelid in its native element, the sea-water; this is repeated at short intervals until death ensues. After the animal has been thus slowly killed, it may be passed through

---

[1] Dr. Mayer first noticed this in objects stained with Kleinenberg's hæmatoxylin, and afterwards in the use of cochineal, where a gray-green precipitate is sometimes produced which renders the preparation worthless. Such results may be avoided by first soaking the objects a few hours in acid alcohol (1–10 parts hydrochloric acid to 100 parts 70 % alcohol).

the different grades of alcohol in the ordinary way, or through other preservative fluids. Objects killed in this manner show no trace of the external crust of precipitates which arises where stronger grades of alcohol are first used. The action of the alcohol is thus moderated, and the animal, dying slowly, remains extended and in such a supple condition that it can easily be placed in any desired position. The violent shock given to animals when thrown alive into alcohol of 40 % to 60 %, giving rise to wrinkles, folds, and distortions of every kind is thus avoided, together with its bad effects.

Acid Alcohol. — In order to avoid the bad effects of alcohol, — such as precipitates, maceration, etc., — Mayer recommends acid alcohol, —

    95 volumes 70 % or 90 % alcohol.
    3   "   hydrochloric acid,[1] —

for larger objects, particularly if they are designed for preservation in museums. The fluid should be frequently shaken up, and the object only allowed to remain until thoroughly saturated, then transferred to pure 70 % or 90 % alcohol, which should be changed a few times in order to remove all traces of the acid. For small and tender objects, acid alcohol, although preferable to pure alcohol, gives less satisfactory results than picro-sulphuric acid.

Boiling Alcohol. — In some cases among the Arthropods, Mayer has found it difficult to kill *immediately* by any of the ordinary means, and for such cases recommends *boiling absolute alcohol*, which kills instantly. For Tracheata this is often the only means by which the dermal tissues can be well preserved, as cold alcohol penetrates too slowly.

Glycerine and Alcohol. —

    Glycerine . . . . 20 parts
    Alcohol (70 %) . . . . 40 "
    Sea water . . . . 40 "

This mixture, poured very slowly into the containing glass, often gives very good results, both for anatomical and histologi-

---

[1] Acid alcohol, as above prepared, loses its original qualities after standing some time, as ether compounds are gradually formed at the expense of the acid.

cal purposes. This method originated with Salvatore Lo Bianco the conservator of the Naples Zoological Station.

Method of Making Absolute Alcohol. — According to Sharp, absolute alcohol is prepared in Ranvier's laboratory by adding anhydrous cupric sulphate to ninety-five % alcohol. Pulverized cupric sulphate is heated to red heat, in order to drive off the water of crystallization; when cool, the white powder is placed in a wide-mouthed bottle, holding about a litre, and three-fourths full of alcohol. The bottle is quickly closed, and the whole shaken. After standing a day or more — with occasional shakings, — it is decanted, and the operation repeated, especially if the cupric sulphate shows much of the blue color due to the reassumption of water. As a test, a drop of the alcohol thus dehydrated may be mixed with a drop of turpentine on a glass slide, and examined under the microscope; if no particles of water are to be seen, the alcohol is absolute enough for all practical purposes.

IODINE.

Iodine, followed by a little strong sulphuric acid, is used as a test for starch and cellulose. In combination with alcohol it has often been used for the central nervous system of the higher vertebrates. It may be strongly recommended in the investigation of many animals, especially transparent Arthropods, where the preparations are required for *immediate* study. The following solution is recommended by Fol: —

To a saturated aqueous solution of iodide of potassium, add as much iodine as will dissolve. This gives a dark-brown fluid, which, according to the end in view, may be diluted with 100–500 parts of water.

Berthold [1] experimented with iodine and other reagents on delicate marine algæ, and came to results that are of importance to zoologists as well as botanists. The aim was to find solutions that would produce the least possible disturbance in the structure of the cell-protoplasm. It was found that satisfactory results could not be obtained with the ordinary

---

[1] Pringsheim's Jahrbücher für Wissenschaftliche Botanik. Vol. XIII. pp. 704-5, 1882.

aqueous solutions of picric acid, osmic acid, etc. The disturbance of the osmotic equilibrium, on transferring delicate cells from sea-water to fresh-water solutions, resulted in intra-cellular derangements. Parallel trials were therefore made of picric acid, osmic acid, and iodine, three different solutions of each being made; one in distilled water, one in alcohol, and another in sea-water. The solutions in distilled water and alcohol proved almost worthless in each case, while each of the solutions in sea-water gave good results. It was found, curiously enough, that the protoplasm of the cells was more easily injured than the nuclei and karyokinetic figures.

Solutions of osmic acid and corrosive sublimate in sea-water gave good preparations, but the iodine solution was regarded as the best reagent. A few drops of a saturated alcoholic solution of iodine, added to the sea-water, gives the desired results. The algæ remain in the solution one half to one minute, and are then transferred directly to 50 % alcohol.

OSMIC ACID.

Osmic acid, one of the most important histological reagents, comes in small closed glass tubes, holding a gram each. It is generally used in very weak solutions ($\frac{1}{10}$–1 %). It is best to prepare a 1 % solution by breaking the tube into a bottle containing 100cc. distilled water. The containing bottle should be double-stoppered (stopper and cap), and should be protected from the light, either by painting black or by inclosure in a cylindrical pasteboard box. From this 1 % solution any desired weaker solution can readily be made. The action of this acid depends on its easy reduction by organic substances.

Objects may be treated in two ways, — either by immersion in the solution itself, or by exposure to its vapor. In the latter case, the object, with little or no water about it, is placed under a bell-jar near a watch-glass containing a small quantity of a strong solution of the acid. The solution does not penetrate more than the superficial portions of large objects; while the vapor reaches even the deepest parts, and produces a very uniform effect.

The time of exposure to the action of the acid should be relatively short (from one to thirty minutes). The blackening of the object may be checked, and removed to a considerable extent, by following the osmic acid with Merkel's fluid or chromic acid.

Mayer employs osmic acid as a staining medium for the hairs, bristles, etc., of the dermal skeleton of Arthropods. The lustre of Sapphirina is preserved by this acid, and, according to Emery, the color of the red and the yellow fatty pigments of fishes.

Van Beneden found osmic acid the best preservative fluid for the Dicyemidæ, and my experience leads to the same conclusion.[1]

Bleaching. — It often happens that objects treated with osmic acid continue to blacken, after removal from the acid, until they are entirely worthless; and such results are even more annoying than the difficulties in the way of staining. It has been said that the blackening process can be arrested by certain staining media; but it is certain that picro-carmine will not always do this, as some of my preparations of Dicyemidæ show. It is therefore a very important step which Dr. Mayer has taken in finding a method of restoring such objects. The method[2] is as follows: —

The objects are placed in 70 or 90 % alcohol, and crystals of potassic chlorate ($KClO_3$) shaken into the liquid until the bottom of the vessel is covered; then a few drops of concentrated hydrochloric acid[3] are added with a pipette, and as soon as chlorine (easily recognized by its greenish-yellow color) begins to be liberated, the whole gently shaken. As soon as the bleaching is finished, the objects are removed to pure alcohol.

---

[1] One of the best objects for testing methods is found in *Phronima sedentaria*. Here the cells and nuclei are so sharply defined that they can be seen in the living animal, and so the effect of a preservative fluid can be easily studied.

[2] A slightly modified form of the method originally given in Mull. Arch., 1874, p. 321.

[3] Nitric acid may be used instead of hydrochloric.

By this method it is possible in half a day to restore large Pelagia, Carinaria, Rhizostoma, etc. Small objects generally require a shorter time and less acid. The process can be greatly accelerated by heating on a water-bath.

Using Sapphirina as a test object, Mayer found that the lustre which characterizes the living animal entirely disappeared by the bleaching process. As this lustre, which has its seat in the epidermis, depends on the interference of light, it is evident that the cells had undergone *some* change, but a change so slight that the tissues could hardly be said to have been injured for histological purposes; besides, the removal of the osmic acid leaves the animal in a good condition for staining.

Mayer's experience with Sapphirina appears to support him in the following conclusions in regard to the nature of the action of osmic acid; viz., that the hardening effect of the acid is due to the formation of inorganic precipitates within the tissues. This is made evident by the fact that the animal becomes soft and flexible as soon as these precipitates are removed by bleaching.

This method of bleaching has been used by Mayer for removing natural pigment. Alcoholic preparations of the eye of Mysis, for instance, can be fully bleached *in toto*, but with better success by operating with single sections. To avoid swelling, which is apt to arise by the use of aqueous fluids, staining media of an alcoholic nature should be used.

ACETIC ACID.

In making solutions of this acid, in all cases where it is important to use a definite strength, it is necessary to use the pure anhydrous form (*acidum aceticum glaciale*), which crystallizes in large, transparent glistening tables that melt at 16°–17° C. Statements respecting its specific gravity cannot be taken as a safe guide to the amount of water contained. It is generally used in solutions of 1 % and 2 %.

Among reagents for animal tissues, this is one of the oldest and best. It is not preservative in its action, and hence is mainly used in investigations where a rapid clearing up of tissues is required, and where temporary preparations alone are sought. It is very useful in studying opaque ova and embryos, and microscopic animals in general. Its usefulness in investigating nuclei is well known.

It has been employed by Kölliker and others in tracing nerves and nerve-terminations, for which purpose Frey recommends —

Glacial acetic acid . . . . 1–2 drops
Water . . . . 100cc.

The action of this acid is rapid, and often requires close watching, in order to catch the most favorable of the transitory conditions which it induces. Recently it has been much employed in combination with other reagents (Flemming, Fol).

CHROMIC ACID.

Chromic acid crystals deliquesce rapidly when exposed to the air, and soon undergo a chemical change. It is therefore important, on opening a bottle of the acid for making a solution, either to dissolve the whole at once, or to close and seal as quickly as possible after taking out the desired quantity.

Solutions are seldom used above 2 % in strength; and this is a very convenient stock solution from which to prepare the more common solutions, ranging from $\frac{1}{4}$ % to 1 %. All chromic solutions are best kept dark, as they undergo changes in the light.

Chromic solutions have, in common with osmic acid, the peculiarity of hardening by virtue of the chemical combinations which they form with cell-substances, and all the consequent disadvantages with respect to staining. The use of chromic acid in the Zoölogical Station of Naples may be said to have been largely superseded by picro-sulphuric acid, corrosive sublimate, and Merkel's fluid, for it is now seldom used except in combination with other fluids. It is sometimes mixed with Kleinenberg's fluid; for example, when a higher degree of hardening is required than can be obtained by the use of the latter fluid alone. It is a common error to use too strong solutions of chromic acid, and to allow them to act too long. Good results are in some cases obtained when the objects are treated with a weak solution ($\frac{1}{4}$ to $\frac{1}{2}$ %), and removed soon after they are completely dead.

Pfitzner has made use of chromic acid, followed by osmic acid, or by chloride of gold, formic acid, and safranin (or hæmatoxylin), for the demonstration of nerve-terminations.

Flemming states that chromic acid is one of the most reliable fixing reagents for the karyokinetic figures, and has proved that objects hardened in this acid can be beautifully and durably stained.

Chromic acid is much used in combination with other reagents (acetic, osmic, picric, nitric acid, etc.).

The more important of these combinations are seen in Perenyi's fluid; the chrom-osmium-acetic acid mixture employed by Flemming, and, in a somewhat modified form, by Fol; the chrom-formic acid used by Rabl in a recent study on karyokinesis; and Merkel's chrom-platinum solution, — recommended by Eisig for annelids, by Brass for Infusoria, and by Whitman for pelagic fish-eggs.

Dr. H. Virchow calls attention to the fact that the precipitates which arise when preparations from any chromic solution (chromic acid, Müller's fluid, etc.) are placed in alcohol may be wholly avoided by keeping in the dark until the alcohol ceases to become colored.

NITRIC ACID.

This is an excellent reagent for many objects. Various strengths have been employed. For chick embryos, His uses 10 %; W. Engelmann found 3-4 % to work well with the retina of higher animals; Strasburger and Flemming used 50 % in studying karyokinetic figures; Rabl-Rückhard used a 10 % solution in tracing the development of the teleostean brain, and Minot finds this solution well adapted for preserving the external form of the brain of the human fœtus.

KLEINENBERG'S PICRO-SULPHURIC ACID.[1]

Picric acid (saturated solution in distilled water) . . . 100 volumes
Sulphuric acid (concentrated) . . . . 2 "

---

[1] Quart. Journ. Mic. Sci., Vol. XIX., p. 208-9, 1879.

Filter the mixture, and dilute it with three times its bulk of water;[1] finally, add as much creosote as will mix.[2]

Objects are left in the fluid three, four, or more hours; and are then, in order to harden and remove the acid, transferred to 70 % alcohol, where they may remain five to six hours. They are next placed in 90 % alcohol, which must be changed at intervals until the yellow tint has wholly disappeared.

The advantages of this fluid are, that it kills quickly, by taking the place of the water of the tissues; that it frees the object from sea-water and the salts contained in it; and that, having done its work, it may be wholly replaced by alcohol. In this latter fact lies the superiority of the fluid over osmic and chromic solutions, all of which produce inorganic precipitates, and thus leave the tissues in a condition unfavorable to staining. Picro-sulphuric acid does not, like chromic solutions, harden the object, but simply kills the cells.

As this fluid penetrates thick chitine with difficulty, it is necessary, in order to obtain good preparations of larger Isopoda, insects, etc., to cut open the body and fill the body-cavity with the liquid by means of a pipette. In larger objects, care should be taken to loosen the internal organs, so that the fluid may find easy access to all parts.

The fluid should be applied as soon as the body is opened, so that the blood may not have time to coagulate and thus bind the organs together. A large quantity of the fluid should be used, and it must be changed as often as it becomes turbid. The same rule holds good in the use of all preservative fluids. It is well also, especially with larger objects, to give the fluid an occasional stirring up.

In order to avoid shrinkage in removing small and tender objects from the acid to the alcohol, it is advisable to take them up by means of a pipette or spatula, so that a few drops

---

[1] Dr. Mayer uses the fluid undiluted for Arthropoda.
[2] Fol gives a more precise formula, —

    Picric acid . . . . 1 g.
    Water . . . . 100 cc.
    Sulphuric acid . . . . 6 cc.

of the acid may be transferred along with them. The objects, sinking quickly to the bottom, remain thus for a short time in the medium with which they are saturated, and are not brought so suddenly into contact with the alcohol. In a few minutes the diffusion is finished; and they may then be placed in a fresh quantity of alcohol, which must be shaken up frequently and renewed from time to time until the acid has been entirely removed.

The sulphuric acid contained in this fluid causes connective tissue to swell, and this fact should be borne in mind in its use with vertebrates. To avoid this difficulty, Kleinenberg has recommended the addition of a few drops of creosote, made from beech-wood tar, to the acid. According to Mayer's experience, however, the addition of creosote makes no perceptible difference in the action of the fluid.[1]

This fluid must not be used with objects (*e. g.*, Echinoderms) possessing calcareous parts which it is desired to preserve, for it dissolves carbonate of lime, and throws it down as crystals of gypsum in the tissues. For such objects picro-nitric acid may be used (Mayer).

PICRO-NITRIC ACID.

Water . . . . 95 parts
Nitric acid (25 % N_2O_5) . . . . 5 "
Picric acid as much as will dissolve.[2]

Picro-nitric acid also dissolves carbonate of lime, but it holds it in solution; and thus the formation of crystals of gypsum is avoided. In the presence of much carbonate of lime, the rapid production of carbonic acid is liable to result in mechanical injury of the tissues; hence in many cases chromic acid is preferable to picro-nitric acid.

Picro-nitric acid is, in most respects, an excellent preservative medium, and, as a rule, will be found to be a good alternative in those cases where picro-sulphuric acid fails to give satisfac-

---

[1] Fol thinks the tendency to swell can be better counteracted by adding to the mixture one third its volume of 1 % chromic acid, leaving out the creosote.

[2] This mixture is used undiluted.

tory results. Mayer commends it very strongly, and states that with eggs containing a large amount of yolk material, like those of Palinurus, it gives better results than nitric, picric, or picro-sulphuric acid. It is not so readily removed from objects as picro-sulphuric acid, and for this reason the latter acid would be used wherever it gives equally good preparations.

PICRO-CHROMIC ACID.

Picric acid (saturated aqueous solution) . . . . 10 parts
Chromic acid (1 %) . . . . 25 "
Water . . . . 65 "

Recommended by Fol[1] as an excellent hardening agent for very small pieces of tissue. It acts slowly, having little power of penetration. A little osmic acid (.005), added shortly before using, is said to strengthen its action much. The staining capacity of objects is not impaired by this mixture. Objects treated with this fluid should be washed in water. The extraction of the acid is more complete and rapid if nearly boiling-hot water is used.

PERENYI'S FLUID.[2]

The following mixture is recommended by Perenyi for hardening eggs and embryos. It kills instantaneously, and preserves the object in its natural form and size without the least shrinkage. Eggs thus preserved are not brittle, but "cut like cartilage."

Nitric acid (10 %) . . . . 4 parts
90 % alcohol . . . . 3 "
Chromic acid (½ %) . . . . 3 "

This mixture assumes a violet hue after a short time.

Eggs may be left in this fluid four to five hours, then twenty-four hours in 70 alcohol, then in 90 alcohol one or two days, finally four to five days in absolute alcohol.

The staining fluid may be best added to the fluid itself, so that the eggs are hardened and colored at the same time. Fuchsine and aniline red dissolve directly in the fluid, while

---

[1] Lehrbuch der vergleich. Anat., p. 100.
[2] Josef Perenyi. "Ueber eine neue Erhärtungsflüssigkeit," Zool. Anzeiger, V., No. 119, p. 459, 1882.

eosine and purpurine are first dissolved in three parts alcohol. Picro-carmine or borax-carmine mixed with the fluid give excellent staining. The addition of staining fluids causes a precipitate, so that, before using, the fluid should be filtered.

For pelagic fish ova I find that ten or fifteen minutes immersion is quite sufficient. My use of this fluid leads me to think that its action should be checked as soon as the object becomes completely permeated by it. It may be followed by Kleinenberg's hæmatoxylin or by Böhm's carmine acetate.

MERKEL'S CHROM-PLATINUM SOLUTION.

Platinum chloride dissolved in water 1:400
Chromic acid   "   " 1:400

Professor Merkel,[1] who employed a mixture of these two solutions in equal parts for the retina, states that he allowed from three to four days for the action of the fluid. Eisig has used this fluid with great success in preparing the delicate lateral organs of the Capitellidæ for sections, and recommends it strongly for other annelids. Eisig allows objects to remain from three to four hours in the fluid, then transfers to 70 % alcohol. With small leeches I have found one hour quite sufficient, and transfer to 50 % alcohol.

For pelagic fish eggs, the best results are obtained by treating first with osmic acid ($\frac{1}{4}$ %) ten minutes, and then with a modified form of Merkel's fluid for two or three days (*vide* methods of preparing fish eggs).

BICHROMATE OF POTASSIUM.

Potassium bichromate acts with less energy than chromic acid, causes very little or no contraction, and for many eggs and embryos,[2] as well as for the central nervous system, is far superior to the simple acid. As a rule, solutions are made stronger and allowed to act much longer than with chromic acid. The more generally used strength is 2 %. Like all chromic solutions, it should be kept dark. The energy of its action is increased by heating; but, in most cases, the better

---

[1] "Ueber die *Macula lutea* des Menschen," etc., Leipzig, 1870, p. 19.
[2] C. Hoffmann has used this fluid extensively for the eggs of invertebrates and fishes.

results are obtained by keeping the fluid in a cool place. A piece of camphor suffices to prevent the appearance of mould.

The process of hardening is so slow that it is often advisable to begin by treatment with some other fluid, and then complete the hardening in this. (See Betz's method of hardening brain and spinal cord.)

Heidenhain (*vide* Special Methods) has introduced a method in which the preparation is both hardened and stained before treatment with bichromate of potassium.

Sectioning should not be delayed long after hardening, as preparations soon become brittle.

MÜLLER'S FLUID.

Distilled water	100 parts
Bichromate of potassium	2 "
Sulphate of sodium	1 "

This excellent fluid was first recommended by H. Müller for the retina, and hence often called 'Müller's Eye-Fluid.' At least two weeks are required for hardening the retina. It is now much employed for hardening the central nervous system (Cf. Weigert, Max Flesch, Osborn, Birge). It hardens very slowly, requiring eight weeks in the case of the vertebrate brain. If kept at a temperature of 30° to 40° C., the process may be completed in eight to ten days (Weigert). Henneguy employed this fluid, after osmic acid, in hardening fish eggs.[1]

Minot makes the following remarks on this fluid: —

In hardening in Müller's Fluid, it is important to keep the jar with specimen and fluid in a cold place — best near freezing — for three or four days. The delicate tissues are then much better preserved. Probably the same is true of Erlicki's fluid. Using Müller's Fluid at a high temperature is bad for epithelia.

ERLICKI'S FLUID.[2]

Distilled water	100 parts
Bichromate of Potassium	2½ "
Cupric sulphate	½ "

---

[1] It has often been highly recommended for delicate embryos; but its value for such purposes has been disputed by Fol.

[2] Progrès médic., Sept. 29, 1877; Revue d. sc. médic., XI., 1, p. 13; and Warschauer med. Zeit., XXIII., Nos. 15 and 18.

Erlicki recommended this fluid for hardening the human brain, for which fourteen days are required. During the first week, the fluid was renewed every day; during the second, once in three days. Thus treated, the brain is well preserved, and hard enough for sectioning.

In Weigert's method, this fluid may replace Müller's with similar results. In Freud's gold chloride method, the oblongata was first treated with Erlicki's fluid.

PERCHLORIDE OF IRON (VULPIAN, FOL).

In experimenting with a delicate class of marine infusoria (Tintinnodea), Fol[1] found that the reagents in common use for instantaneous killing, — such as picro-sulphuric acid, osmic acid, alone or in combination with chromic and acetic acid, and corrosive sublimate, — failed to give successful preparations. He finally succeeded with perchloride of iron, a reagent quite new in histological technique. An alcoholic solution diluted to about 2 % will answer ordinary purposes; but a stronger solution should be used in case it is desired to kill a large number of animals in a large vessel. It will not do, however, to turn a saturated solution directly into sea-water, as precipitates would be copiously formed, which would utterly ruin the preparations. After the animals have sunk to the bottom of the vessel, most of the water may be turned off, and 70 % alcohol added. In order to remove from the tissues the ferric salts adhering to them, it is necessary to replace this alcohol with alcohol containing a few drops of hydrochloric acid.[2]

For general use, the following solution is recommended:[3] —

Perchloride of iron (alcoholic solution) . . . . . 1 part
Alcohol (70 %) . . . . 5–10 "

---

[1] Zeitschr. f. wiss. Zool. XXXVIII., p. 491, 1883.

[2] The best method of staining such objects is to add a few drops of gallic acid (one per cent solution) to the alcohol. After twenty-four hours the acidulated alcohol is turned off, and pure alcohol added. Thus treated, the protoplasm will take a light brown color, the nuclei a much deeper brown. Carmine stains too deeply and diffusely, and cannot be successfully removed.

[3] Fol, Lehrbuch, p. 102.

Ciliated epithelium is exposed to this solution a few minutes, then washed in 50 % alcohol, to which $\frac{1}{2}$ to 1 % oxalic acid has been added; and finally, after the complete disappearance of the yellowish-red color, transferred for preservation to pure alcohol.

CORROSIVE SUBLIMATE.

The saturated aqueous solution, now most generally employed, is prepared by heating the water (with excess of sublimate) to boiling. The cold solution contains only about two per cent of the salt. A three or four per cent solution can be made in 50 to 60 % alcohol.

This reagent kills, in most cases, instantly, and leaves the tissues in a good condition for staining and sectioning. It is important, however, to cut at an early date, as objects thus killed become brittle if kept long in alcohol.

Sublimate has been brought into general use through Dr. Lang's recommendation. Prompted by a statement found in an old paper by Blanchard,[1] he began experimenting with corrosive sublimate as a medium for killing marine planarians; and his marked success led him and others to employ the same with other animals. In most cases, Dr. Lang now uses a saturated solution of corrosive sublimate in water. A saturated solution in picro-sulphuric acid, which, in some cases, gives better results if a little acetic acid (five per cent or less) is added, is also used.[2] Blanchard's mode of treatment was to mix a quantity of the aqueous solution with the sea-water, and thus poison the animals.

Lang, on the contrary, removes the sea-water so far as possi-

---

[1] "Recherches sur l'Organisation des Vers," by Emile Blanchard. Ann. des Sci. Nat. Zool. Ser. 3, t. VIII., 1847, p. 247.

[2] These solutions given in Zoolog. Anzeiger, 1879, II., p. 46. The original solution (Zoolog. Anzeiger, 1878, I., p. 14–15), now little used, stood thus: —

Distilled water	100 parts
Common salt	6–10 "
Acetic acid	5–8 "
Corrosive sublimate	3–12 "
Alum (in some cases)	$\frac{1}{2}$ "

ble before applying the solution. With planarians he proceeds in the following manner:—

The animal is laid on its back, and the water removed with a pipette; the solution being then poured over it, it dies quickly and remains extended. After half an hour, it is washed by placing it in water and changing the water several times during thirty minutes. It is next passed through 50 %, 70 %, 90 %, and 100 % alcohol. In two days it is fully hardened, and should then be stained and imbedded in paraffin as early as possible, *as it is liable to become brittle if left long in alcohol.* The time required by the corrosive sublimate varies with different objects, according to size and the character of the tissues. As a general rule, it may be said that objects should be removed from the fluid as soon as they have become thoroughly saturated by it. In order to kill more quickly than can sometimes be done at the ordinary temperature, the solution is heated, and in very difficult cases may be used boiling.

Corrosive sublimate has been used with success by Dr. Lang and others in the following cases: Hydroids, Corals, Nemertines, Gephyrea, Balanoglossus, Echinoderms, Sagitta, Annelids, Rhabdocœla, Dendrocœla, Cestodes, Trematodes, embryos and adult tissues of Vertebrates, and, according to Mayer and Giesbrecht, Crustacea with thin chitinous envelopes, *e. g.*, Sapphirina, Copepods, and larvæ of Decapods.

The two great advantages of Dr. Lang's method are, (1) that animals so treated are easily stained, and (2) that they are killed so quickly that they are left, in most cases, in a fully extended condition. Hot corrosive sublimate kills leeches so instantaneously that they often remain in the attitude assumed the moment before the fluid is poured over them. The color, however, is not so well preserved as when killed with alcohol, or even with weak chromic acid.

It should be remembered that objects lying in a solution of corrosive sublimate *must not be touched with iron or steel instruments;* wood, glass, or platinum may be used.

FLUIDS FOR MACERATION.

The separation of tissues and tissue elements may be accomplished in a great variety of ways. It may be effected through *mechanical* processes (by the aid of dissecting needles, by shaking, and by pressure), or by *chemical* means (macerating agents etc.). The methods and means must be varied to suit the nature of the object under study, and according to the purpose in view. It may be desirable merely to *dissociate* certain elements; or it may be required to *destroy* some parts while preserving others.

Most of the fluids in common use for killing and hardening may be converted into macerating and dissociative agencies by diluting to various degrees with water. Water alone is often a valuable means of maceration. Weak alcohol (30 % or less), acetic acid (1 % or less), dilute osmic acid (alone, or in combination with other fluids), chromic acid ($\frac{1}{2}-\frac{1}{10}$ %), Müller's fluid, and iodized serum are among the milder agents of dissociation; while *eau de javelle*, hydrochloric acid, and caustic potash belong to the more violent. The following will serve as illustrations.

ANDRES' METHOD WITH ACTINIÆ.

It is often important to see the cells of a tissue *in situ* before freeing them with needles. In such cases Dr. Andres proceeds as follows: —

1. Killed with corrosive sublimate.
2. Left in 25 % alcohol twenty-four hours.
3. Soaked for a short time in a very thin solution of gum arabic, then in a somewhat thicker solution, and finally imbedded in a very thick solution.
4. Hardened in 90 % alcohol.
5. Thick sections prepared for dissection with needles. The sections are placed on a slide in water, which dissolves the gum.

THE HERTWIGS' MACERATING FLUID.[1]

For the isolation of tissues in the Cœlenterates, Oscar and Richard Hertwig recommend the following mixture: —

Acetic acid	1 part
Osmic acid	$\frac{1}{4}$ "
Sea water	1000 "

By means of this fluid, not only the nerve cells, muscle cells, etc., can be isolated, so that the exact form of the individual cells may be easily recognized, but also the tissues, in the form of thin lamellæ, may be separated and studied as a whole. Pieces of tissue or whole animals are left in the mixture five to ten minutes, and then washed for several hours in $\frac{1}{4}$ % acetic acid. The macerated parts can be further prepared and afterwards colored on the slide; or they can be colored at once, before preparation with needles. In the first case, picro-carmine is used; in the second, Beale's carmine, because it does not harden the tissues, but assists rather the process of maceration. Pieces of tissue may be preserved a long time in glycerine diluted with an equal volume of water, provided a few drops of carbolic acid have been added to secure against mould and Bacteria.

To obtain preparations of single-cell elements of Actiniæ, the macerated portion must be carefully divided up into smaller parts by needles, and one or more of these parts placed under the cover-glass. Light blows on the cover-glass with a needle will cause the cells to separate. Care should be taken to support one side of the cover by a hair, which is removed quite gradually after the object has been reduced to very small cell masses. Sliding of the cover may be avoided by placing wax feet under its corners.

Dr. Mark has employed this method, and obtained excellent results with it. As he remarks, the great merit of this fluid is, that it separates the cell elements and hardens them at the same time. The *dissociative* and the *preservative* agent are combined in such proportions that the action of the former is confined within desired limits by that of the latter.

---

[1] Jenaische Zeitschrift, XIII., p. 462, 1879.

IODIZED SERUM.

Tincture of iodine	1g.
Amniotic fluid	100g.
Carbolic acid	1–2 drops

Schäfer (Practical Histology, p. 45) prepares by adding to the amniotic fluid "a crystal or two of iodine, and allowing it to stand for a few days with frequent agitation, and then filtering off the clear fluid from any precipitate that may have formed. Almost any other fresh serous fluid can be used instead, but the amniotic fluid is generally the most readily obtained in sufficient quantity, and, at the same time, perfectly clear and pellucid. The serous fluid serves to macerate tissues which are immersed in it, whilst preserving as nearly as possible the original form of their elements. The purpose of the iodine is to prevent putrefaction; it serves also, at the same time, to render the tissue elements slightly firmer."

Ranvier (Traité Technique, p. 76) recommends keeping a *strongly iodized* serum in stock, to be used for iodizing fresh serum. From one to two months are required to prepare the stock solution. Iodine dissolves very slowly at first; but after a time (fifteen to twenty days) more will dissolve, and at the end of two months a very strong solution of a dark brown color is obtained. Fresh serum may then be prepared for use by the addition of a little of this strong solution, — just enough to give it a light brown color.

A piece of tissue may be kept in the weak serum for several weeks without danger of putrefaction, provided that *a few drops of the stock solution is added as often as the brown color disappears.* The fading of the color is due to the absorption of the iodine by the object. "The addition of new quantities of iodine," says Ranvier, "is the key to this method of dissociation."

EAU DE JAVELLE (KClO).

Perls,[1] Altmann, and Noll[2] recommend *Eau de Javelle* as an excellent fluid for removing the soft parts of animal and vege-

---

[1] Arch. f. mikr. Anat., vol. XVI., 1879.
[2] Zoolog. Anzeiger, V., No. 122, p. 528.

table tissues. If a piece of Spongilla, for example, is placed on a slide, and a drop of this water added, all the soft parts are destroyed in from twenty to thirty minutes, and the spicula are left *in situ*. After the protoplasmic parts have been thus removed, the preparation is carefully treated with acetic acid, in order to remove any cloudy precipitates, then washed with weak, strong, and absolute alcohol successively, and finally mounted in balsam.

Very neat preparations of diatoms may be obtained with this fluid. The calcareous part of shells, thus treated, is not destroyed. Sections of plant buds were successfully treated, and then mounted in Meyer's fluid (ten volumes glycerine, twenty distilled water, and three salicyl-pyroligneous acid).

The skeletons of small animals may be easily prepared by placing the bodies in Eau de Javelle, which removes the skin, muscles, &c., without injuring the bones.

CAUSTIC POTASH.

Strong caustic potash (30%) is much better than weak solutions, as it destroys the tissue elements less rapidly. Solutions should be kept in well-stoppered bottles, and renewed after some months, as they readily take up carbonic acid from the air, and change into carbonate of potassium.

DECALCIFICATION.

Decalcification of bone-tissue is best accomplished by the action of weak chromic acid, a reagent which preserves and hardens the included soft tissues at the same time that it dissolves the earthy matter. Its decalcifying power is much weaker than that of hydrochloric acid; but the latter is destructive to soft tissues. The action of chromic acid may be both improved and intensified by combination with nitric acid, in the following proportions:—

Chromic acid . . . . 70 parts
Nitric acid . . . . 3 "
Water . . . . 200 "

When chromic acid alone is used, it is well to begin with a solution of about $\frac{1}{5}$ %, using a large quantity of the fluid and renewing it often, and then follow with stronger solutions, rising gradually to ½ % as recommended by Schäfer.

In using picric acid, the solution should be a saturated one; and fresh crystals of the acid must be frequently added to take the place of that which is used up in dissolving the lime salts.

DESILICIFICATION.

The silicious skeletons of animals (Radiolaria, Fibrospongiæ) may be removed, according to P. Mayer,[1] with fluorhydric acid, without injury to the tissues.

The objects are first hardened in alcohol, and then the acid is added in drops. The duration of the process will vary from a few minutes to a day, according to the size of the object. The operation should take place in the open air, in vessels made of lead, silver, or gutta percha. In the employment of this acid, the manipulator must take special care to avoid its poisonous fumes. The corrosive power of the acid makes it highly dangerous; contact with the skin produces wounds difficult to heal.

---

[1] Zool. Anzeiger, No. 97, pp. 592-593.

CHAPTER II.

METHODS OF STAINING.

For staining purposes, alcoholic solutions are generally to be preferred to aqueous media. The disadvantages in the use of the latter, in staining alcoholic preparations, — such as the tearing asunder of fragile tissues, caused by the violent osmosis, swelling, the effects of which cannot always be fully obliterated by again transferring to alcohol, and maceration, which is liable to result where objects are left for a considerable time in the staining liquid, — may all be avoided by using alcoholic solutions. Objects once successfully hardened may be left in such solutions for any required time, and, when sufficiently stained, be washed in alcohol of a corresponding strength, and then passed through the higher grades without being exposed to water from first to last. As a rule, alcoholic dyes work quickly, and give far more satisfactory results than can be obtained with other media. They penetrate objects more readily, and thus give a more uniform coloring where objects are immersed *in toto*. Even chitinous envelopes are seldom able to prevent the action of these fluids.

It is not, however, to be denied that non-alcoholic dyes may often do excellent work, and, in certain cases, even better than can be otherwise obtained. The swelling caused by aqueous staining fluids is not always an evil, but precisely what is required by some objects after particular methods of treatment.

From experiments recently made, Mayer has found that dyes containing a high percentage of alcohol stain more diffusely than those of weaker grades, from which he infers that strong alcohol robs, to a certain extent, the tissues of their selective power, and renders them more or less equally receptive of coloring matter.

CARMINE SOLUTIONS.

CARMINIC ACID.[1]

Since Gerlach first called attention to carmine as a coloring matter for histological purposes, many methods of preparing the carmine solutions have been published, but, so far as I am aware, none which result in anything but mixtures of various carmine salts with undetermined and undesirable impurities, such as fatty matter, tyrosin, sugar, and salts of alkaline metals. Of course, the coloration produced by these mixtures has been sometimes good, sometimes bad, and the solutions have easily spoiled or decomposed, while their preparation often required considerable time and trouble. Several years ago I studied carmine, chiefly from a chemical standpoint, and since then I have often wondered why naturalists usually used carmine solutions in which water, with some caustic or destructive material added, was the principal solvent. Carmine of commerce, it is true, is not readily soluble, even in water, until ammonia, borax, or some other aid to solution is added; but carminic acid, the basis of the coloring matter of carmine, has long been stated, in the leading chemical dictionaries and handbooks, to be readily soluble in water and in alcohol. Watts (Dict. Chem., 1872, 1st suppl., p. 413) says of carminic acid: "This acid forms a purple mass, fusible and soluble in all proportions in water and in alcohol. Sulphuric and hydrochloric acid dissolve it without alteration. It bears a heat of $136°$ C. without decomposition."

Earlier still, Watts (Dict. Chem., 1863, vol. I., p. 804) says: "The fine red pigment known in commerce as carmine is pre-

---

[1] Geo. Dimmock. Read at the New York meeting of the Society of Naturalists of the Eastern United States, 28th Dec., 1883. — Am. Nat. XVIII., p. 324, 1884.

pared by treating a solution of cochineal with cream of tartar, alum, or acid oxalate of potassium. The fatty and albuminous matters then coagulate and carry down the coloring matter with them." Now in preparing most carmine solutions, this precipitation takes place; and the carmine, having greater cohesive (not chemical) affinity for impurities of animal origin than for alcohol, its solution is not readily accomplished by that medium, nor, indeed, by water. In preparing carmine solutions for histological purposes by some of the published recipes, more than one half of the coloring matter of the carmine is lost in the refuse left upon the filter paper.

There are two ways commonly in use for preparing carminic acid, which term I here use as distinguished from "acid carmine" (the cochineal decoction for which Grenacher, Schweigger-Seidel, and others have given recipes), and from carmine (the impure commercial powder). The first mode of preparation, the one which I followed in preparing the acid upon which I have experimented, is that of Warren De la Rue. Watts, whom I quote because he gives clear directions (Dict. Chem., 1863, vol. I., p. 804) gives De la Rue's method, as follows: "To separate it [carminic acid], cochineal is exhausted with boiling water; the extract is precipitated by subacetate of lead slightly acidulated, care being taken not to add the lead solution in excess; the precipitate is washed with distilled water till the wash-water no longer gives a precipitate with a solution of mercuric chloride, then decomposed by sulphuretted hydrogen; the filtrate is evaporated to a syrupy consistence, and dried over the water-bath; and the dark purple product thus obtained is treated with alcohol, which extracts the carminic acid." The second mode of preparing carminic acid, that of Schaller, is given by Watts (Dict. Chem., 1872, 1st suppl., p. 413), as follows: "C. Schaller prepares this acid by precipitating the aqueous extract of cochineal with neutral lead acetate, slightly acidulated with acetic acid; decomposing the washed precipitate with sulphuric acid, again precipitating the filtrate with lead acetate, and decomposing the precipitate with sulphuric acid, avoiding an excess; then precipitating a third time, and decomposing the precipitate with hydrogen sulphide.

The filtered solution is evaporated to dryness; the residue dissolves in absolute alcohol; the crystalline nodules of carminic acid, obtained on leaving this solution to evaporate, are freed from a yellow substance by washing with cold water, which dissolves only the carminic acid; and the residue left on evaporating the aqueous solution is recrystallized from absolute alcohol or from ether."

Schaller's mode of preparation gives purer carminic acid than De la Rue's; but either kind is sufficiently pure for histological purposes. The precipitation by lead acetate and the dissolving in alcohol free the carminic acid from animal impurities, and the consequence is a purer form of pigment than can be extracted by any process hitherto employed for the preparation of carmine for histological purposes. I will here add that carminic acid has been shown recently by Liebermann and Van Dorp to be related, through nitrococcusic acid, to trinitrocresol, which has been obtained from coal-tar cresol, thus showing that the coloring matter of cochineal contains methylated benzine-residues.

I do not know whether carminic acid, ready prepared, is obtainable. Any chemical student could easily prepare it by the processes given above.

It is unnecessary to explain to naturalists the advantages of alcoholic solutions of carmine over aqueous ones. The alcoholic solution colors preparations much quicker than the aqueous solution does; for coloring sections, I employ a solution of 0.25 gram carminic acid to 100 grams of eighty per cent alcohol, and leave sections in the solution from two to five minutes. A solution of equal carmine strength, but in absolute alcohol, can be employed; it has, however, no special advantages, since, with the eighty per cent alcoholic solution, the sections can be washed directly in absolute alcohol, and then put into oil of cloves or turpentine. Coloring in the piece before sectioning never takes as long with alcoholic carminic acid as it does with ordinary carmine solutions; and, if it did take long, the strong alcohol would preserve the tissue from maceration. In coloring pieces of Mollusca, or of other equally slimy animals, the slime should be removed beforehand, or the coloration will be unsatisfactory, because the slime, congealing in the alcohol, takes up the coloring matter, forming an almost impervious colored layer on the outside, and leaving the inside of the piece nearly uncolored.

Some preparations colored in alcoholic carminic acid, and then put up in glycerine, lost their color in a few months, the color seeming to be entirely diffused in the glycerine, while similar preparations mounted in Canada balsam retained their color perfectly. I do not know if this fading would occur with preparations colored with alcoholic ammonic carminate, or even if this diffusion was not due to some impurity of the glycerine (of the purity of which I was doubtful). Time to test this matter further failed.

An alcoholic ammonic carminate, or ammonia carmine, can be prepared, at a moment's notice, from alcoholic carminic acid, by adding ammonia, drop by drop, and stirring until the entire solution changes from its bright red to purple red. By this mode, pure alcoholic ammonic carbonate can be produced, with no excess of ammonia, and at any time. As the carminic acid can be preserved dry without decomposition, and dissolves quickly in alcohol, one can carry the ingredients of a carmine solution in the vest pocket without inconvenience.

In making and using alcoholic carminic acid, pure alcohol and distilled water give the best results, because a portion of the carminic acid is converted to carminates by the salts of impure water. In making alcoholic ammonic carminate, this precaution is not as necessary, because the color of the carminates produced by the impurities of the water is so nearly like that of ammonic carminate.

Alcoholic carminic acid may be used, as Grenacher's carmine solution is used, to color sections from which the coloration is to be afterwards partly extracted by very dilute hydrochloric acid, leaving nuclei red. Another way to use carmine solutions, which is especially applicable to alcoholic carminic acid, is to precipitate the carmine in the tissues by some salt, the carminate of the base of which gives a desired coloration. I have found, for example, that specimens hardened for a moment under the cover-glass with an alcoholic solution of corrosive sublimate (mercuric chloride) and, after washing with alcohol, colored in alcoholic carminic acid, took a fine coloration of mercuric carminate. So, too, specimens colored in alcoholic carminic acid can be changed by a few moments' treatment with a *very dilute* alcoholic solution of lead acetate or cobalt nitrate to a beautiful purple. With lead acetate, used as above, a double coloration is sometimes produced; but I have not examined these colorations sufficiently to describe them accurately. Cupric and other salts, used as above described, have not given me very favorable results. Sometimes salts in the tissues of the animals themselves change portions of the carminic acid to purple carminates, giving a double coloration without further treatment.

Picric acid added to alcoholic carminic acid *in extremely small quantities* (best in a dilute alcoholic solution, testing the solution on specimens after each addition) makes a double alcoholic coloring fluid (a so-called picro-carmine). I have been unable, thus far, to determine the proportion of picric acid required for this solution, having in every case added an excess. All different kinds of carmine solutions can be made from carminic acid, with the advantage of having always uniform strength, of being definite mixtures, and of not spoiling as readily as those made directly from cochineal.

Incompatible reagents with carminic acid are, of course, all alkaline solutions, and nearly all metallic salts; with ammonic carminate, are naturally all acids; with all carmine solutions, are bromine and chlorine.

BEALE'S CARMINE.

Carmine . . . . 1 gram
Ammonia . . . . 3 cc.
Pure glycerine . . . . 96 cc.
Distilled water . . . . 96 cc.
Alcohol, 95 % . . . . 24 cc.

Dissolve in the ammonia plus part of the water; add the rest of the water, and allow to stand in an open dish in a warm place until the ammonia is nearly all driven off. Then add the alcohol and glycerine.

For use, dilute with an equal part of glycerine, which must be thoroughly mixed in by stirring. Stain in an open dish, which, together with a second open dish containing acetic acid, is placed under a bell glass or in a closed box. The staining requires at least twenty-four hours, and much longer if the carmine is fresh. Beale's carmine improves with use, and what is left, after being employed as directed, should be filtered back into the original bottle.

From the carmine solution the sections are placed in water and washed *thoroughly*, after which they are placed for from one to three minutes in hydrochloric acid, diluted with water until it tastes about like sharp vinegar. They are finally again washed in water and are then ready to mount in the usual manner.

Employed in this way, Beale's carmine is one of the most valuable of histological staining fluids, both for general use and also more especially for the central nervous system. If by any chance the sections are over-stained, the superfluous color may be extracted by a *brief* sojourn in *very dilute* ammonia — one drop of strong ammonia to five cubic centimetres of water is quite sufficient; this is to be followed by a rapid washing in water, and an immediate transfer to dilute hydrochloric acid. (Minot).

ALCOHOL CARMINE (MAYER).[1]

Carmine	4 g.
Alcohol (80 %)	100 cc.
Hydrochloric acid	30 drops

The pulverized carmine is added to the acidified alcohol, and the mixture boiled for thirty minutes, to dissolve the coloring substance. The solution is filtered warm, neutralized by adding ammonia until a precipitate begins to appear. After cooling, it may be necessary to filter several times.

GRENACHER'S CARMINE SOLUTIONS.[2]

(1) **Alum Carmine.** — An aqueous solution of alum (1 to 5 %, or any degree of concentration) boiled with $\frac{1}{2}$ to 1 % pow-

---

[1] Gravis. Bull. de la Soc. Belg. de Micr., 1884, No. VII., p. 126.

[2] Arch. f. mikr. Anat., Vol. XVI., p. 463, 1879. None of these solutions to be used where calcareous parts are to be preserved.

dered carmine for 10 to 20 minutes; allowed to cool, then filtered. With the addition of a little carbolic acid the fluid will keep for years. It colors quickly, and nuclei more strongly than other parts. Objects washed in water after staining.

(2) **Acid Borax Carmine.** — *a.* An aqueous solution of borax (1 to 2 %) and carmine ($\frac{1}{2}$ to $\frac{3}{4}$ %) heated till the carmine is dissolved.

b. Acetic acid added by drops to solution *a*, while shaking, until the color is about the same as that of Beale's carmine.

c. Solution *b* left standing 24 hours, then turned off and filtered.

This solution, which is a modification of Schweigger-Seidel's acid carmine, is not recommended for coloring *in toto*. It colors sections in one half to three minutes diffusely, and hence, after washing in water, they are placed for a few minutes in alcohol (50 % or 70 %) to which a drop of hydrochloric acid has been added; then transferred to pure alcohol.

(3) **Borax Carmine.**[1] — *a.* An aqueous solution of borax (4 %) and carmine, heated till the carmine is dissolved.

b. Solution *a* mixed with 70 % alcohol in equal parts, left standing 24 hours and filtered.

This fluid may be used for coloring objects in toto. After staining, the objects are to be washed in 35 % alcohol, to which a little hydrochloric acid has been added (4 to 6 drops to 100 cc.), and allowed to remain here until the color has been sufficiently removed. They are next passed through successively higher grades of alcohol for hardening.

(4) **Alcohol Carmine.** — A teaspoonful of carmine dissolved, by heating about 10 minutes in 50 cc. of 60 to 80 % alcohol, to which 3 to 4 drops of hydrochloric acid have been added, then filtered.

Objects colored in this fluid should not be washed in water, but in alcohol of a grade corresponding to that of the solution.

---

[1] Dr. Mayer prepares, for some purposes, borax carmine of 50, 60 or 70%. That of 70% contains little carmine, but is well adapted to staining delicate objects that would suffer if exposed to weaker solutions. Boiling alcohol (50% or 60%) dissolves about 1% carmine and 1% borax.

For diluting alcoholic solutions of carmine, alcohol of the same strength must always be used.

RANVIER'S PICRO-CARMINATE.[1]

1. Ranvier takes a saturated solution of picric acid, and adds to it a saturated solution of carmine in ammonia (until a cloudy precipitate begins to appear).

2. This solution is then evaporated by heat until reduced to about one-fifth of its original volume, then cooled and filtered.

3. The filtered solution is then evaporated to dryness, leaving a reddish powder, which, as often as occasion requires, is dissolved in distilled water (1 : 100).

In using this picro-carminate, one can have the double stain of the carmine and the acid; or the simple stain of the carmine; the acid being washed out by soaking in distilled water.

The solution, when properly prepared, should be neutral. If the solution obtained from the powder is not neutral, it can be made so by adding a little carbonate of ammonia and heating moderately in an open vessel, as recommended by Fol. Ammonia or carbonic acid escapes, and the solution becomes at once neutral.

PERGENS' PICROCARMINE.[2]

I. 1. Boil, for $2\frac{1}{2}$ hours, 500 g. of cochineal (pulverized) in 30 litres of water.

2. Add 50 g. potassic nitrate, and, after a moment of boiling, 60 g. oxalate of potash; boil 15 minutes.

3. After cooling, the carmine settles: it is washed several times with distilled water, in the course of three or four weeks.

II. 4. Pour a mixture of 1 volume of ammonia with 4 volumes of water upon the carmine, taking care that the carmine remains in excess.

5. After two days, filter, and leave the filtered solution exposed to the air until a precipitate forms.

6. Filter again, and add a concentrated solution of picric acid; agitate, and then allow it to stand 24 hours.

7. Filter, and add 1 g. chloral for one litre of the liquid.

---

[1] Traité Technique, p. 100.
[2] Biologie Cellulaire, by J. B. Carnoy, p. 92, 1884.

8. At the end of eight days separate the liquid from the slight precipitate which is formed, and it is ready for use.

This fluid keeps unchanged for at least two years, and is recommended by Carnoy above other picro-carmine solutions.

MINOT'S PICRIC-ACID CARMINE.

Boil one gram best powdered carmine with 200 cc. of water, plus an excess of picric acid for half an hour; allow it to stand and cool; decant the clear fluid, add fresh water, and, if necessary, picric acid; boil, cool, and decant; repeat this operation until all the carmine is dissolved. Place the decanted fluid in an evaporating dish, add about one gram thymol, and stand in a warm place until the volume is reduced to 25 cc.; let the solution cool, filter, wash out the residue, which should be on the filter, with 25 cc. water; dilute the filtrate with 50 cc. water. By this means, a solution ready for use, which will keep indefinitely, and contains carmine and picric acid in good proportions, can be prepared with certainty.

It gives a stronger differential coloring than Ranvier's picro-carmine; but over-staining must be most carefully avoided. For staining sections, two to five minutes are sufficient. The fluid stains connective tissue (fibrous) deep red; striped muscle, deep, dull red; smooth muscle, blood, and horny tissue, bright yellow; glands, reddish yellow. With the kidney it gives differentiation of the different portions of the tubules; for the central nervous system it seems to be of little value. If rightly used, it gives a sharp, nuclear coloring.

If the aqueous solution is evaporated to dryness, the residue may be redissolved in alcohol, giving an alcoholic carmine dye. This I have not yet tested sufficiently. Apparently the alcoholic solution will keep only a few months. The alcoholic solubility of the dye offers the advantage that sections stained in the watery solution can be washed in alcohol directly.

PICRO-CARMINE.

A very excellent picro-carmine is prepared by Mayer in the following manner: —

To a mixture of powdered carmine (2 g.) with water (25

cc.), while heating over a water-bath, add sufficient ammonia to dissolve the carmine. The solution may then be left open for a few weeks, in order that the ammonia may evaporate; or the evaporation may be accelerated by heating. So long as any ammonia remains, large bubbles will form while boiling; but as soon as the free ammonia has been expelled, the bubbles will be small, and the color of the fluid begin to be a little lighter. It is then allowed to cool, and filtered. To the filtered solution is added a concentrated aqueous solution of picric acid (about four volumes of the acid to one of the carmine solution).[1]

A small crystal of thymol, or a little chloral-hydrate (1 % or more), will serve as an antiseptic.

ACETIC ACID CARMINE.[2]

Pulverized carmine added to a small quantity of boiling acetic acid (45 %) until no more will dissolve; filtered, and diluted to about 1 % for use.

Flemming used the concentrated solution.

BÖHM'S CARMINE ACETATE.

1. Carmine (4 g.) pulverized in 200 cc. water.
2. Ammonia added by drops until the solution becomes cherry-red (the carmine should now be fully dissolved).
3. Acetic acid slowly added until the cherry-red color becomes brick-red. The addition of acetic acid should be accompanied with stirring, and should cease the moment the change in color is effected.
4. Filter until no trace of a precipitate remains.

If the color is not sufficiently deep, a few drops of ammonia should be added before filtering, and the solution left in an open vessel until the alkali has volatilized.

Objects may be left for twenty-four hours, or more, in this fluid. The deep stain should be partially removed by immer-

---

[1] The addition of the acid should cease before a precipitate begins to form.

[2] Schneider. Zool. Anzeiger, No. 56, p. 254, 1880.

sion in a mixture of water (50 volumes), glycerine (50 volumes), and muriatic acid (½ volume), for a few minutes. The karyokinetic figures are thus brought out with great distinctness.

HÆMATOXYLIN AND COCHINEAL.

GRENACHER'S HÆMATOXYLIN.

Saturated alcoholic solution of hæmatoxylin . . 4 cc.
Strong aqueous solution of ammonia-alum . . . 150 "

Let this mixture stand 8 days in the light; then filter and add, —

Glycerine . . . . 25 cc.
Methylalcohol . . . . 25 "

This fluid works best after standing a few months, or, at least, long enough for a sediment to form at the bottom.

KLEINENBERG'S HÆMATOXYLIN.[1]

1. To a saturated solution of chloride of calcium[2] in 70 % alcohol add a little alum, and filter.

2. One volume of No. 1 mixed with six to eight volumes of 70 % alcohol.

3. At time of using, pour into No. 2 as many drops of a concentrated solution of crystallized hæmatoxylin in absolute alcohol as suffice to give the required depth of color.[3]

---

[1] May be used after all hardening fluids.

[2] Chloride of calcium, according to Kleinenberg, has no other use than to strengthen the osmotic action between the hæmatoxylin solution and the alcohol contained in the tissues. As chloride of calcium and alum give a precipitate of gypsum, it would probably be better to use chloride of aluminum.

[3] A good solution should be violet, inclining a little to blue. The red tinge that arises after the fluid has stood for some time indicates that it has become slightly acid, in which condition it is unfit for use. To restore its proper color, it is only necessary to open a bottle of ammonia over the mouth of the bottle holding the hæmatoxylin, in such a manner that a very small quantity of the gas will mix with the fluid. If too much ammonia gas be added, a precipitate is produced which spoils the fluid.

If the color appears too strong, the fluid may be diluted with solution No. 1.

Before immersing objects in this fluid, great care should be taken to free them from the least trace of acid by frequently changing the alcohol. If this is not done thoroughly, the acid left in the preparation will sooner or later cause the color to fade; and such results have led to the erroneous conclusion that hæmatoxylin will not give durable preparations.

Small objects are best stained in a weak solution, which colors more slowly, but with greater clearness, than stronger solutions. After staining, Kleinenberg transfers objects to 90 % alcohol. In case of over-staining, the color may be partly removed by adding a little oxalic acid or hydrochloric acid ($\frac{1}{2}$ %, or less) to the alcohol containing the objects. The acidulated alcohol is allowed to work until the color is slightly reddened. On transferring to pure alcohol, the color passes again into a permanent blue-violet.

MAYER'S COCHINEAL TINCTURE.

One gram powdered cochineal soaked in 8 to 10 cc. of 70 % alcohol for several days, then filtered.

The clear, deep-red fluid thus prepared may, like hæmatoxylin, be used in all cases where it is desirable to stain with an alcoholic solution, and will be found particularly useful for objects that are not easily penetrated by the ordinary aqueous solutions of carmine, such as the Arthropods.

It is necessary, before immersing larger objects in this fluid, to leave them a short time in 70 % alcohol, otherwise there may be a precipitate. The time required for staining will vary from a few minutes to a day or more, according to the nature and size of the object. With larger objects, requiring considerable time, it is important to use a large quantity of the fluid, otherwise the amount of coloring stuff in solution might not suffice to give the proper depth of color. Small and delicate objects, on the other hand, may be most successfully treated with a solution which has been diluted with 70 % alcohol, or one which has been weakened by previous use. It is always necessary to free the tissues, after staining, from the surplus dye; and this may be done by washing in 70 % alcohol, which must be changed until it shows no color. This process requires, for larger objects, considerable time and alcohol, but may be hastened by using the alcohol slightly warm.

The color ultimately assumed by objects treated with cochineal tincture varies much, and depends partly on the reaction of the tissues themselves, partly on the presence or absence of certain salts. It is certainly one of the best recommendations of this staining agent that, varying with the nature of the object and its mode of treatment both before and after staining, it gives such an extraordinary diversity of results. On account of the great variety of substances contained in the dried dye-stuff, it is evident that the composition of the tincture must vary according to the strength of the alcohol employed as a solvent. Solutions in 90 % or 100 % alcohol have a light-red color, and stain too diffusely to have any practical value. The weaker the alcohol the stronger the tincture, and the stronger the alcohol the more easily it penetrates objects; the grade of alcohol may therefore be selected with reference to two points, depth of color and readiness of penetration; 70 % or 60 % is recommended by Mayer as combining both these qualities in a very favorable degree. It is important to remember that, whatever be the strength of the solution, a precipitate will always be produced if an alcohol of a different grade, whether higher or lower, be mixed with it. It is evident, then, that a tincture of any given strength contains substances that are insoluble in any other grade of alcohol, and this explains why superfluous coloring matter can only be removed from objects by the aid of alcohol of precisely the same degree as that of the tincture.

Overstaining, which seldom occurs, may be easily corrected by the aid of acid alcohol ($\frac{1}{10}$ % hydrochloric acid, or 1 % acetic acid). Acid makes the tincture lighter, more yellowish-red, while the addition of ammonia and other caustic alkalies changes it to deep purple. Still more important is the fact that salts soluble in alcohol give a blue-gray, green-gray or blue-black precipitate. For example, if a piece of cloth that has been dyed in cochineal, and washed, be treated with an alcoholic solution of a ferric or a calcic salt, it will assume a more or less deep blue color.

As the salts present in the living organism are seldom, if ever, fully removed by preservative fluids, but in some cases even increased, it will often happen that an object, though stained in the red fluid, comes out blue, precisely as when stained with hæmatoxylin. Such a result cannot, however, be obtained in the presence of acids, nor in the absence of inorganic salts; under these conditions the color is always red. It is not possible, therefore, to know what color an object will ultimately present.

Very often the different tissues of one and the same object present unlike colors. In the embryos of Lumbricus, Kleinenberg found the walls of the blood-vessels red, their contents dark blue. Glandular tissues, or their contents, are frequently stained gray-green.

Objects treated with chromic or picric solutions, or with alcohol, usually stain without difficulty; but osmic acid preparations should be bleached before staining. Cochineal does not color so intensely as hæmatoxylin, and hence the latter often gives more satisfactory results in the case of large objects stained in toto.

As before pointed out, alcohol causes the salts contained in sea water to be precipitated, thus forming a crust on the exterior of the animal which interferes with the staining process. It is therefore necessary to treat marine animals that have been preserved in strong alcohol, with acid alcohol (1–10 parts hydrochloric acid to 1,000 parts 70 % alcohol), and then carefully wash in pure 70 % alcohol before staining with cochineal.

ANILINE DYES.

As a rule, aniline colors cannot be employed for staining objects in toto.

With very small objects and sections already cut, very excellent results can be obtained by the methods developed by Böttcher,[1] Hermann,[2] Flemming,[3] and others; for here diffuse stain-

---

[1] Müller's Archiv, 1869, p. 373. Virchow's Archiv, Bd. XL., p. 302.
[2] Communicated to the Naturforscherversammlung in Graz, 1875. Tagblatt, p. 105.
[3] Archiv f. mikr. Anat., Bd. XIII. p. 702, Bd. XVI. p. 302, Bd. XVIII. p. 151, Bd. XIX. p. 317, and p. 745, Bd. XX. p. 1.

ing may generally be avoided by first overstaining and then withdrawing the color to any desired extent by means of alcohol. But to obtain satisfactory results, the sections must be thin enough to allow uniformity of action both to the coloring and the decoloring agent. It is evident that the process cannot be similarly controlled in larger objects, particularly where a dye is used, which, like most of those under consideration, is quickly extracted by alcohol, for in this case the color would be removed from the superficial layers more rapidly than from the deeper ones, so that a uniform precision of color would be impossible.

BISMARCK BROWN.[1]

Bismarck brown forms an exception to the statement made in the last paragraph.

A saturated solution is made by dissolving the powder in boiling water or weak alcohol, or, according to Mayer, in 70 % alcohol. The solution should be used undiluted, and requires to be filtered from time to time. It colors very quickly objects hardened in alcohol or chromic acid.

EOSIN.

The best strength for section-staining is 0.0005. Dissolve one half gram eosin in one litre of 95 % alcohol — or dissolve one gram in 100 cc. alcohol, and for use dilute one part of the stock solution with 20 parts alcohol. The alcoholic solution is far more convenient than the aqueous. (Minot.)

ROSE BENGALE IN COMBINATION WITH IODINE GREEN AND BLEU DE LYON.

Rose bengale, according to Griesbach,[2] is the bluest of the eosin dyes. An aqueous solution is very useful in staining chromic acid preparations of the spinal cord, as it colors the gray substance much more strongly than the white substance.

Rose bengale may be used in combination with iodine green and bleu de lyon in the following manner:

---

[1] Weigert. Arch. f. mikr. Anat., XV. p. 258, 1878.
[2] Zool. Anzeiger VI., No. 135, p. 172.

A section of an alcoholic preparation is first placed in distilled water; then quickly drawn through a deep red solution of rose bengale, and again placed in aq. dest. It is next laid in iodine green for a few seconds, washed, and placed for five minutes in absolute alcohol to remove any excess of color. It may be transferred directly from the alcohol to a dilute alcoholic solution of bleu de lyon (two parts alcohol abs. and three parts aqua dest.) and left for a few seconds, then replaced in the absolute alcohol preparatory to mounting.

SAFRANIN.

One part safranin dissolved in 100 parts of absolute alcohol; after a few days 200 parts of distilled water is added.

Pfitzner,[1] from whom the above formula is taken, recommends this solution as one of the best for staining nuclei. It is cheap, easily prepared, acts quickly, and stains *only* the nuclei. It works best with chromic acid preparations, from which the acid has been removed as much as possible.

NIGROSIN COMBINED WITH PICRIC ACID.[2]

To a saturated aqueous solution of picric acid, add a small quantity of an aqueous solution of nigrosin. This gives a deep olive-green mixture which kills very quickly, and stains at the same time it hardens. After a few hours immersion the object may be transferred to water or alcohol for removing the acid and the nigrosin which remains in solution.

This method is very good for preparing small objects that have to be killed, hardened, stained, &c., under the cover-glass.

DAHLIA (Monophenyl-rosanilin).[3]

Dahlia is soluble in alcohol (one sort is soluble in water). The aqueous solution stains the protoplasm blue-violet, leaving the nuclei very pale. If sections thus stained are treated with very dilute acetic acid, the protoplasm is decolored and the nuclei

---

[1] Pfitzner. Morph. Jahrb., VI. pp. 478–480 and VII. p. 291.
[2] Pfitzner. Ber. Dtsch. Botan. Gessellsch., 1, No. 1, p. 44, 1883.
[3] Ehrlich. Arch. f. mikr. Anat., XIII., pp. 263–277.

become blue-violet. The staining fluid is best prepared as follows: —

Alcohol (absolute)	50 cc.
Distilled water	100 cc.
Glacial acetic acid	12½ cc.
Dahlia (to near saturation).	

The dye should be allowed to act at least twelve hours.

BLUE BLACK.

Dissolves easily in water, with difficulty in alcohol.

Blue black	½ g., dissolved in
Aq. dest.	1 to 2 cc. ; then add
Alcohol	99 cc.

Recommended by Sankey,[1] especially for the central nervous system.

MYRTILLUS.

Dr. M. Lavdowsky[2] recommends the juice of fresh huckleberries, *Vaccinium* (*Gaylussacia* Gray) *myrtillus*, as an excellent staining medium, especially for the karyokinetic figures and the cellulose walls of plant cells.

Preparation. — The newly picked berries should be washed in water, then the juice expressed and mixed with twice its volume of distilled water containing a few grams of 90 % alcohol. The mixture may next be boiled for a short time, then filtered while it is still warm. The fluid thus obtained is deep red and slightly acid, and may be kept in well-corked bottles for a long time if not left in too warm a place. The fluid becomes somewhat thick, on cooling, and may therefore be diluted with two to three times its volume of distilled water at the time of using.

Two different colors are produced by this fluid: 1. A *red color* is produced by the action of the fluid on fresh objects, and on objects hardened in chromic acid or other chromic solutions. The best staining is said to be obtained from chromic acid preparations. 2. A *lilac color* — more durable than the

---

[1] Quart. Jour. Mic. Sci., 1876, p. 35.
[2] Arch. f. mikr. Anat., XXIII., pp. 506–508.

red — may be obtained in the following manner : — Place three watch-glasses on a white ground, fill one with the myrtillus fluid, the second with a 1 % solution of acetate of lead, and the third with distilled water. Place the object in the dye for one or two minutes, wash in the distilled water, then leave in the solution of acetate of lead until the lilac color becomes pronounced; finally, wash and mount in glycerine, or, after treatment with alcohol, in balsam. In case glycerine be used, a little acetate of lead should be mixed with it.

DOUBLE STAINING WITH ANILINE DYES.[1]

The method of experiment adopted by Vincent Harris was as follows : — The blood was spread in thin layers upon cover-glasses, and allowed to dry in direct sunlight. The dried blood was then wet with a few drops of some dye, and, after five minutes, washed with a slow stream of water from a wash-bottle; it was then dried in the flame of a spirit-lamp, and allowed to cool. A second dye was then applied in the same way; and, after washing, the preparation was mounted in balsam, without having recourse to alcohol and clove oil.

The following combinations of dyes were found to give good results : —

Rosein and iodine green.
Fuchsin and methylen blue.
Fuchsin and Bismarck brown.
Eosin and vesuvin.[2]
Iodine green and Bismarck brown.
Hoffman's violet and Bismarck brown.
Methyl violet and methylen blue.

Rosein, followed by iodine green, stained the colored corpuscles a bright red, with bluish-green nuclei; and the colorless corpuscles were so stained that three varieties could be readily distinguished.

Fuchsin and methylen formed a very successful combination. The methylen blue was used as a saturated solution in absolute alcohol.

---

[1] Quart. Jour. Mic. Sci., XXIII., p. 292.
[2] Regarded as identical with Bismarck brown.

Bismarck brown was prepared as a 2 % solution in dilute alcohol. In the use of this dye it was found best to immerse the preparation twenty to thirty hours, as the color then remains, even when passed through alcohol and clove oil.

An aqueous solution of Hoffman's violet was used with a dilute spirit solution of Bismarck brown.

The green dyes are not permanent. The solutions should be quite fresh in order to secure successful results.

THE DIFFERENTIAL ACTION OF SAFRANIN AND METHYL GREEN.

In studying the sexual characteristics of the oyster, Ryder[1] found that a mixture of these two dyes enabled him to distinguish both ova and spermatozoa in the same follicle, the nuclei of the ova being stained red by the safranin, and the heads of the spermatozoa bluish-green by the methyl green. The method of preparation is as follows: —

1. After removing the shell, the oyster is hardened in chromic acid (1 to 2 %) for several days.

2. Washed in water 2 days, and then further hardened in alcohol.

3. Soaked for 24 hours in water, to remove the alcohol; then imbedded in gum arabic, and cut with free hand.

4. Sections freed from imbedding mass by washing in water; then stained in a mixture in equal parts of

        Safranin   (saturated alcoholic solution)
        Methyl Green   "   "   "

diluted with eight times its volume of water, 2 to 3 hours.

5. Decolored in 95 % alcohol until clouds of color no longer appear (5 to 15 minutes).

6. Clarified in clove oil and mounted in balsam or dammar.

STAINING INTRA VITAM.

Bismarck Brown. — Karl Brandt,[2] who was the first to recommend Bismarck brown for coloring living organisms,

---

[1] Bull. U. S. Fish. Com., 1883.

[2] Sitzgsber. Berl. physiol. Gesellsch., pp. 34, 35, Dec. 1878; and Biolog. Centralblatt, p. 202, 1881.

found that, by using a weak solution (1 : 3000 or 1 : 5000) on living cells (fresh-water Protozoa, etc.), the nuclei remained uncolored, while fatty substances were stained brown. He succeeded in coloring the nuclei with a dilute aqueous solution of hæmatoxylin, the organisms still remaining alive.

Paul Mayer[1] mixed Bismarck brown with sea-water in which Caprellidæ were kept, and found that the contents of the stomach and the "secretion-balls" of the liver (hepato-pancreas) were soon colored brown; and that, after a somewhat longer time, the muscles were stained diffusely, and the blood-corpuscles showed one or more brown dots. When the specimens thus treated were placed in pure sea-water, the stain wholly disappeared within a few hours, leaving them quite uninjured.

Cyanin (Bleu de Quinoléine, Chinolinblau). — A. Certes[2] recommends a weak aqueous solution of cyanin (1 : 500,000 - 1 : 100,000) for staining Infusoria alive. The coloring substance is not completely soluble in water. The solution should be made with ordinary water, filtered, as distilled water is poisonous to Infusoria. Infusoria may be kept alive in the solution from twenty-four to thirty-six hours. The fatty granulations of the protoplasm stain deeply; the protoplasm, cuticle, and cilia very feebly; and the nuclei still less, or not at all. Solutions of cyanin, whether in water or alcohol, are decolored more or less rapidly by the light, and should therefore be kept dark.

Methyl and Gentian Violet. — J. Bizzozero[3] employed the two following mixtures in the study of blood corpuscles: —

Methyl violet (concentrated aqueous solution) . .	1 part
Salt solution ($\frac{1}{2}$ %)	5000 "
Gentian violet (saturated solution)	1 part
Salt solution ($\frac{1}{2}$ %)	3000 "

---

[1] Fauna u. Flora des Golfes von Neapel. VI. Monographie: Caprelliden. p. 153, 1882.

[2] Zool. Anzeiger, IV., p. 208 and 287, 1881.

[3] Arch. f. pathol. Anat., p. 261, 1882.

CLASSIFIED LIST OF THE CHIEF ANILIN DYES, WITH THEIR SOLUBILITIES IN WATER AND SPIRIT (VINCENT HARRIS).

BROWN.	RED.	ORANGE.	YELLOW.	GREEN.	BLUE.	VIOLET.
Bismarck — Partially sol. in water; sol. in dilute spirit.	*Eosin* — Pink; freely sol. in water.	*Aurin* — Insol. in water; partly sol. in strong spirit; more so in absolute alcohol.	*Fluorescin* — Greenish-yellow; insol. in water; sol. in spirit, the solution being beautifully fluorescent.	*Iodine Green* — Blue green; freely sol. in water or spirit.	*Soluble Anilin Blue* — Freely sol. in water.	*Hoffman's Violet* — Freely sol. in water and in dilute spirit.
Vesuvin — Sol. in water.	*Anilin Scarlet* — Insol. in water; sol. in methylated spirit.	*Anilin Orange* — Ditto, ditto.	*Anilin Primrose* — Only partly sol. in meth. spirits.	*Malachite Green*, a less bine-green; freely sol. in water and in spirit.	*Bleu de Lyon* — Insol. in water; freely sol. in strong spirit.	*Methyl Violet* — The red predominating; sol. in water partially; freely sol. in water.
Chrysoidin — Sol. in water.	*Flamingo* — Deep brownish-red; partly sol. in water; freely so in meth. spirit.	*Tropeolin* — In deep yellow glistening scales; partly sol. in water; more sol. in methyl. spirit.			*Methylen Blue* — A very deep blue; freely sol. in water and in spirit.	*Gentian Violet* — The blue predominating; freely sol. in water.
	Ponceau¹ — Deep red-crimson; partly sol. in water; freely so in dilute spirit.	*Phosphin* — Yellowish orange; partly sol. in water; more so in spirit.			*China Blue* — Freely sol. in water.	*Tyrian Blue* — Near to violet; sol. in water.
	Rosanilin — Partly sol. in water; freely sol. in dilute spirit.	*Safranin* — Sol. in water and in spirit.			*Serge Blue* — Ditto.	*Spiller's Purple* — Sol. in spirit.
	Fuchsin — Partly sol. in water; sol. in dilute spirit.				*Blue Black* — Freely sol. in water.	

¹ Ponceau is a mixture of rosaniline and phosphin.

FILTERING REAGENT BOTTLE.[1]

A wide-mouth bottle of convenient size is fitted with a cork, through which three glass tubes are passed, one of which (Fig. 1, *a*) reaches nearly to the bottom of the bottle, while the other two tubes extend only just below the cork.

The long tube is curved above the cork; and at a point just beyond the curve there is attached to it a short piece of tubing (*b*) of twice the diameter of the first. The lower end of this larger tube (at *d*) is also fitted with a cork, through which passes a short piece of small tubing, which is slightly contracted at its distal end. Over the cork and open end of this short delivery tube, a piece of fine muslin (*d*) is stretched, and the space (*b*) in the large tube is filled with a loosely packed plug of absorbent cotton, forming an effectual filter. The second tube ends above the cork in a "thistle-bulb" funnel (*e*), the opening in which is formed into a neck of sufficient length to enable one to cork it securely. Through this thistle-top tube the reagent may be readily poured from a dish or other vessel.

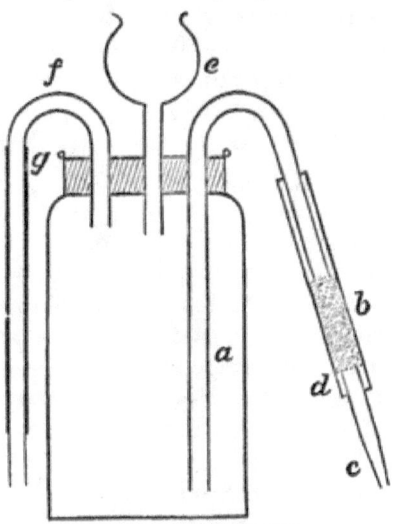

FIG. 1.— Filtering bottle.

The third tube (*g*) is bent over the side of the bottle, and a piece of rubber tubing is attached, which may end either in a short glass mouth-piece or atomizer-bulb. In case the bulb is used, it will be found necessary to loosen the cork in the funnel-tube (*e*) when a sufficient quantity of the contents has been expelled, in order to release the pressure and stop the flow.

When the plug of cotton becomes foul, which will not be for

---

[1] C. E. Hanaman. Am. Mo. Mic. Journ., March, 1883.

a long time, even with hæmatoxylin stain, it may be replaced by a fresh one by simply uncorking the lower end (*d*) of the large tube, pulling out the old plug with a pair of forceps, and inserting a fresh one. The cotton should be but loosely packed to work easily.

IMPREGNATION WITH GOLD CHLORIDE.

Gold chloride is one of the best, and at the same time one of the most variable, agents for staining nerve fibres. Its uncertain behavior is unquestionably due to imperfections in the methods of using it; and these imperfections are attributable in part to circumstances and conditions beyond control, but mainly to a failure to recognize some details in the process which are essential to success. The discoverer of a new method of using this agent is rarely able to claim that its application is attended with uniformly successful results, even in his own experience; and still less is he able to prescribe rules which shall guide others to invariable success. A capital illustration of the unreliableness of gold-chloride methods in general is furnished in that of Prof. Flechsig ("Die Leitungsbahnen im Gehirn und Rückenmark des Menschen," 1876). After stating a method that has conducted him to successful results, Flechsig adds, "We are not yet in a position to give directions with sufficient exactness to guarantee certain success in the operation, since the conditions which work especially favorably or unfavorably are not precisely known." Of this method, as of most others, the best that can be said is, that it is very valuable if it succeeds. (*Vide* Chapter XIII.)

METHODS OF BLEACHING OVER-COLORED GOLD PREPARATIONS.

Redding[1] washes the over-stained tissues in weak ferricyanide of potassium.

Cybulsky[2] decolored with cyanide of potassium. The object

---

[1] Proc. Amer. Micr. Soc., p. 183, 1882.
[2] Zeitschr. f. wiss. Zool., XXXIX., p. 657, 658, 1883.

is placed in a few drops of a $\frac{1}{2}$ % solution on the slide, and allowed to remain from one half to two minutes. As soon as the color has faded sufficiently, the solution is removed by the aid of filtering paper, replaced with glycerine, and then covered with a cover-glass. By this process the object is not only partially bleached, but the epithelia are loosened and easily removed by pressure on the cover.

IMPREGNATION WITH NITRATE OF SILVER.

Nitrate of silver[1] is usually employed in solutions of $\frac{1}{2}$–2%, but may also be used in a solid form. In the latter form, it is applicable only to dense tissues, such as the cornea and cartilage; and the layers with which it comes in direct contact are destroyed.[2] In the case of the cornea, the solid piece is passed rapidly over the anterior surface of the membrane while still *in situ*. The cornea is then detached and laid in distilled water, and the epithelium removed with a brush. The nitrate of silver dissolves in the water that bathes the object, passes through the epithelium, and is reduced in the fibrous tissue by the action of the light.[3]

Solutions are employed for two purposes: first, to bring out clearly cell boundaries in epithelia; second, to demonstrate small cavities and lymph-spaces. In the first case, through the impregnation of the intercellular substance, so-called negative images are obtained; while in the second, positive ones are produced, the cavities and spaces themselves becoming black.

Pieces of fresh epithelium may be left in the solution a few minutes (from 5 to 30), then transferred to a 2 % solution of acetic acid, and exposed to strong daylight or direct sunlight until they turn distinctly brown.

The over-blackening of the preparations may be prevented by washing them in a 10 % solution of hyposulphite of sodium ($Na_2S_2O_3$), as recommended by Legros.[4] The silver compound

---

[1] Kept in black bottles.
[2] Heitzmann, Micr. Morph. of the Animal Body, p. 9.
[3] Ranvier. Traité Technique, p. 105.
[4] Journ. de l'Anatomie, 1868, No. 3, p. 275.

(insoluble in water), which has not been acted upon by the light, forms with the hyposulphite a double salt, which is soluble in water; and this double salt, together with the superfluous hyposulphite, is removed by washing the preparation thoroughly in distilled water.

Bleaching. — Preparations that have become too black may be bleached by the action of the hyposulphite, or by cyanide of potassium (1 : 10).

In the case of marine animals, a difficulty in the way of using nitrate of silver is met with, for salt precipitates the silver instantly. This difficulty may be avoided in two ways: first, the object may be killed in weak osmic acid, and then, after washing in distilled water long enough to remove most of the salt, placed in the silver solution, and treated in the ordinary way;[1] second, the object may be transferred from sea water to a 0.5 % solution of potassic nitrate, in which the tissues are freed from the greater part of their chlorides in a short time, and from which, while still living, the object is transferred to the silver solution. (v. HARMER'S METHOD.)

Alcoholic preparations may be treated with an alcoholic solution of silver nitrate (*circa* 2 %)[2].

In order to demonstrate the existence of small cavities in the tissues ("positive images" — His), it is necessary to expose the object to a rather strong solution for a longer time than usual (in the dark). It is next placed in a solution of salt (2–5%) for a few minutes, and then washed in distilled water, and left for a time in the light. The spaces appear to be filled with black granules.

In the investigation of nerve-fibres, Golgi has combined treatment with a mixture of bichromate of potassium and osmic acid with silver impregnation. (v. Chapter XIII.)

PALLADIUM CHLORIDE (Schulze).

F. E. Schulze,[3] to whom we are indebted for the introduction of this agent into histological technique, recommended it as an

---

[1] R. Hertwig. Jen. Zeitschr. f. Nat., XIV., pp. 313, 324.
[2] Fol. Lehrbuch d. vergl. mik. Anat., p. 179.
[3] Centralbl. f. d. med. Wiss., No. 13, 1867.

excellent means of hardening and staining small pieces (size of a bean or smaller) of organs, especially for muscular tissue. Muscle-fibres, gland-cells, and epithelia stain bright yellow, while fatty and connective tissue remain uncolored.

Henle and Merkel[1] employed it for sections of the central nervous system, for which purpose it has since been almost exclusively used in Germany.

The sections are first placed in a $\frac{1}{8}-\frac{1}{2}\%$ palladium chloride for one to two minutes, then transferred directly to a strong solution of ammonia carmine (a minute or more). Thus treated, the axis-cylinder stains bright red, the medullary sheath yellow.

Thanhoffer recommends palladium dichloride for the nerves of the cornea. Palladium has been very highly praised by Bastian, and through him it has acquired general recognition in England.

Schulze used a $\frac{1}{10}\%$ solution, and allowed it to act twenty-four hours, at the end of which time the preparation is hardened sufficiently for sectioning. According to Fol (op. cit., p. 180), smooth muscular fibres are stained clear yellow, while striated fibres take a brownish hue.

---

[1] Henle's Handbuch der Nervenlehre der Menschen, 1871.

CHAPTER III.

INSTRUMENTS.

MICROTOMES.

Perhaps no instrument of scientific research has ever undergone more rapid alteration and variation than has the microtome within the last ten years, and especially during the last five. The rapid multiplication of different patterns, the endless number of improvements and additions, all show how very important, as an instrument of investigation, the microtome has become in the estimation of embryologists, histologists, anatomists, and physiologists. The sneers that greeted its introduction from some quarters have hastily subsided; and now it is universally acknowledged to be an indispensable part of the outfit of every respectable laboratory.

The various microtomes now in use may all be regarded as modifications of two types (Oschatz and Rivet). The chief difference between the two lies in the method of raising the object. In one of these — the best models of which are the Schanze microtome (Leipzig) and the Zeiss microtome — the object is raised by a micrometer screw; while in the sledge microtome — the best form of which is the Thoma instrument made by Jung of Heidelberg — it is raised by being pushed up an inclined plane. In this difference lies the superiority of the sledge over the screw microtome. A finer precision and a more perfect uniformity in the elevation of the object are attainable with the latter than with the former. It is hardly possible to reach a fineness of $\frac{1}{500}^{mm}$ in the vertical elevation of the object by means of the screw; while in the sledge principle of construction the possibility of raising the object only $\frac{1}{1000}^{mm}$ has already been realized.

An interesting combination of the sliding and screw movement is presented in the new microtome made by C. Reichert,[1] in Vienna. In this instrument the object is raised automatically, which must be regarded as an important improvement.

As the Thoma, Schanze, and Caldwell microtomes are now most widely known and used, they have been selected for description.

THE THOMA MICROTOME.

This microtome (Fig. 2) consists of a stand of cast iron, on which slide two carriers. The large knife is attached to one of

FIG. 2.—Thoma's microtome. *a*, carrier for the knife; *b*, carrier for the object; *c*, micrometer-screw for fine adjustment.

these, *a*, which slides horizontally. The second, *b*, holds the specimen to be cut. This second moves on an inclined surface, so as to raise the specimen as required.

This, with a few modifications, is the general character of all sliding microtomes; but hitherto the carriers were constructed to slide with two even surfaces between two even planes of the stand, which intersect at a given angle, with the consequence that all show more or less imperfect results, owing to the fact that it is impossible to obtain sufficiently exact plane surfaces.

---

[1] Described by Moeller in Zeitschr. f. wiss. Mikroskopie, I., Heft 2, pp. 241-244, 1884.

The inconveniences appear in small, scarcely perceptible irregularities of the movement of the carriers, and the consequent impossibility of making sections as thin as with an experienced hand.

This induced Professor Thoma to enter upon a consideration of the geometrical and mechanical difficulties to be surmounted. The question to be solved was, how many points at least of a body sliding between two planes must touch the latter for this body to be perfectly steady in its position. It will be found that five points are sufficient, and that a carrier on five points, between two plane surfaces, will slide without difficulty between these planes, even if they are not absolutely geometrical planes, or the angle which they include is not everywhere the same. Such a carrier will always take exactly the same course; and, in consequence, a knife attached to it will cut a series of perfectly parallel sections through an object which is successively raised to a higher plane after each cut. The working of the instrument will therefore be far superior to any microtome with large sliding surfaces which nowhere exactly fit the sliding surfaces of the stand. This indicates the desirability of constructing the carrier for the object also on five points.

The construction resulting from these principles is simple and practical, but it is necessary to take into consideration the centres of gravity of the different sliding bodies. This, however, complicates the matter but very little. We replace the two sliding surfaces of each carrier by five slightly prominent points, and they will then move with exactness on any combination of two planes not differing too much from geometrically plane surfaces. One condition only must be fulfilled, namely, that the five points are so chosen as to support steadily the centre of gravity of the carriers, including their accessory parts, namely, the knife and object. Fig. 3 gives a more precise idea of the details of construction.

In the figure, the lower surfaces of the carrier, a, which supports the knife, show three prominences, which give the geometrical projection of the five points. Within the limits of the figures these points could not be drawn exactly as they are in the instrument itself. In reality, they appear only as small prominences upon three narrow ridges on the sliding surfaces of the stand. This arrangement was desirable to facilitate the action of the oil with which the sliding surfaces are to be covered. Two of the ridges form together parts of the oblique plane, and the third corresponds to the vertical sliding plane. The same arrangement is found in the carrier, b, which supports the clamp in which the object is placed.

By this mode of construction the carriers will move gently and regularly, even if the sliding surfaces on the stand are not perfect geometrical planes. It is still, however, of course, desirable that as much exactness as possible should be obtained in these planes, as their irregularities cannot fail to affect the sections, especially as they are, in fact, multiplied in the latter.

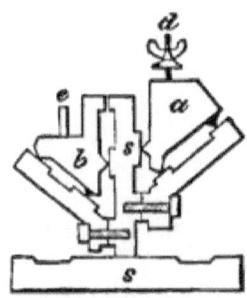

FIG. 3.—Transverse section of the microtome. ss, stand; a, carrier for the knife; b, carrier for the object.

Specimens which are well hardened will allow of sections of three to four square centimeters surface and 0.015 to 0.10^{mm} thickness. In exceptional cases pieces of so large a surface may be cut of 0.005^{mm} thickness. If the section is smaller (for instance, one centimeter square), the thickness can be reduced considerably, say to 0.005^{mm}, or in extreme cases to half that. It is not, however, all tissues and objects that will admit of sections of such delicacy. Well-hardened liver may generally be cut to 0.015^{mm}, this being about the diameter of the hardened cell. Occasionally, however, in this tissue, sections of 0.010^{mm} can be obtained. Lymphatic glands and brain may be cut to 0.010 or 0.075^{mm}; embryonic tissues, well imbedded, usually admit sections of 0.005 and 0.003^{mm}. In some cases even sections of 0.002^{mm} thickness can be obtained. These numbers refer to the largest size of the microtome, and to serial sections. The two smaller sizes will give sections of the same delicacy, but comparatively smaller in extent of surface. The length of the sliding surfaces of the large instrument is 40^{cm}, and the edge of the knife is 23^{cm}. In the medium size these dimensions are 27 and 16^{cm}, and in the smallest about 21 and 11^{cm}.

Sliding microtomes are in general constructed for cutting sections of tissues previously hardened in alcohol, picric acid, chromic salts, and other agents. Fresh tissues are decidedly better cut by freezing microtomes — for instance, on the simple and practical instrument of Hughes and Lewis. The addition of a freezing apparatus to a thoroughly exact sliding microtome is neither advisable nor necessary. The differences of temperature produced in different parts of the instrument would be apt to interfere with the perfect planeness of the sliding surfaces; whilst, on the other hand, section-cutting with frozen tissues is so simple and easy with the ordinary freezing apparatus, that any further complication in the way of a sliding support of the knife is superfluous.

In cutting, the microtome is to be placed before the operator as in Fig. 2, with the sliding surfaces abundantly covered with oil (bone oil). In many cases it will be sufficient to simply place the hardened specimen between the arms of the clamp attached to the carrier, *b*. The clamp should then be fixed in such a position that the specimen is as near as possible to the knife-carrier. The knife will generally have to be adjusted so as to bring the whole length of its blade into action. Very hard specimens are frequently cut with less difficulty by placing the knife more obliquely in regard to the long diameter of the instrument.

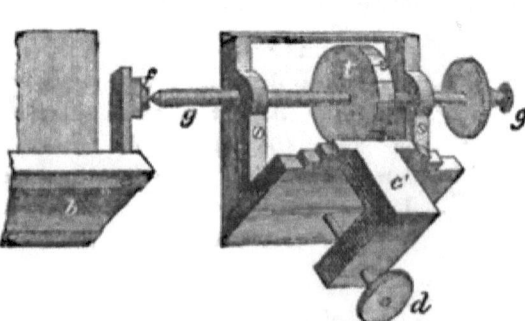

FIG. 4.—Micrometer-screw for delicate sections.

The inclination of the oblique plane upon which the carrier, *b*, slides is 1 : 20, and consequently the section will be 1-20th^{mm} thick if the carrier is moved 1^{mm} on the oblique plane. A scale in millimeters with a vernier allows the operations to be exactly regulated. The vernier will be found sufficient for sections of 0.015^{mm}. Sections of greater delicacy should always be made by using the micrometer-screw, which was designed to obtain the utmost exactitude in the management of the carrier, b. Fig. 4 shows it on a larger scale.

The carrier, c' slides on the same oblique plane as the carrier b which holds the specimen. In all positions of the latter it is therefore possible to bring the point of the micrometer screw, gg, close to a small polished plate of agate, f, which is fixed to the carrier, b. In this position c should be firmly screwed to the stand of the microtome by d, and every revolution of the micrometer-screw, gg, will then push the carrier, b, 0.3^{mm}. The periphery of the drum, t, which is firmly attached to the screw, gg, is divided into fifteen equal parts; and consequently each division marks a thickness of section equivalent to 0.001^{mm}. The finest sections hitherto produced reach only 0.002^{mm} thickness.[1]

Since the first microtome was taken into use, a series of important improvements have been made. One of them consists in a clamp (Fig. 5) for holding the object, which can be turned round three axes, and admits, therefore, of a very easy adjustment of the object in regard to the knife. It was devised to meet the desire for occasionally turning the object between two successive series of sections.

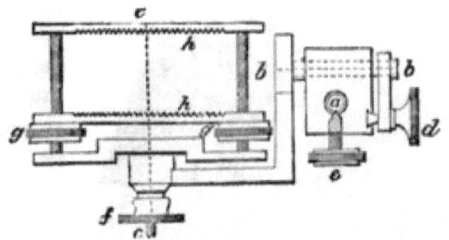

FIG. 5.— Clamp to be turned in three directions (as seen from above).

The two metal plates, hh, form the jaws of the clamp. Between them is placed the cork which carries the specimen, and

---

[1] Professor R. Kossman writes: "Many to whom the turning back of the micrometer-screw of the microtome is an annoying delay will be thankful to me for pointing out to them that in two or three seconds it can be turned back its whole length by using a kind of fiddle-bow, such as is used for drilling holes. The loop of the bow-string (made of strong silk cord, waxed or rosined) is passed round the smooth neck of the screw, and the bow is moved alternately to the left with stretched, and to the right with slackened, cord.

the latter is fixed by turning the screws, $g\,g$. The three axes are a, $b\,b$, and $c\,c$, and round these the clamp can be turned, a being vertical, and $b\,b$ and $c\,c$ horizontal. In all positions these three axes can be made immovable by the screws d, e, and f. The axis, a, is formed by the vertical rod, e (Fig. 3), on the carrier supporting the clamp and object. The details of the construction are partly new, and are very solid and durable. Their arrangement is such as to admit of a division of the circles in which the clamp can be turned. Another improvement is the registering micrometer-screw described below.

The Registering Micrometer-Screw. — In the improved form of the micrometer-screw, invented and described by Mayer, Andres, and Giesbrecht, an arrangement for regulating its movement has been added. It consists of a spring which, after a given number of divisions of the drum, registers to the ear and finger of the manipulator the number of micromillimeters which the object has been raised. The intervals between the registering clicks can be varied by means of a vernier-like adjustment of the two halves of the drum, so as to equal an entire revolution of the drum, or only $\frac{1}{15}$, $\frac{1}{3}$, $\frac{1}{2}$ of a revolution.

Fig. 6.—Drum of the new micrometer-screw.

An examination of Fig. 6, which illustrates the new form of the drum, will show how the intervals are regulated. The drum is composed of two symmetrical halves, $A\,B$ and $A'\,B'$, so closely apposed that the dividing line (dotted in the figure) is scarcely visible. The periphery of each half is composed of two zones of unequal radii. The larger zones, B and B', are in apposition, and together form the graduated portion of the drum. Each of the smaller zones are marked with the figures 1, 2, 3, and 15. When the drum is in order for work, it rotates with the screw, which is marked gg in Fig. 4.

The left half of the drum, AB, is held in position by the screw, s, and may be rotated independently of the right half, $A'B'$, or of the screw, gg, by the aid of a handle which fits the holes xxx.

When the half AB is adjusted to the half $A'B'$, in the manner represented in the figure, the fifteen equal parts into which the zone B is divided exactly correspond to the same number of parts in the zone B', so that the grooves which mark these parts in one zone become continuous with those of the other zone. Thus adjusted, the spring, which rides on the ruffle, BB', with a sharp edge parallel to the grooves, will give fifteen sharp clicks in the course of one rotation of the drum, the click being heard every time the sharp edge falls into coincident grooves. In order to adjust for fifteen clicks, it is only necessary to rotate AB until groove fifteen becomes continuous with groove fifteen of the opposite half ($A'B'$). For one click in one rotation, the grooves 1 1 must be made to coincide; for two clicks, the grooves 2 2; and for three clicks, the grooves 3 3. The intervals between successive clicks may thus be made to correspond to $\frac{1}{1}$, $\frac{1}{2}$, $\frac{1}{3}$, or $\frac{1}{15}$ of a complete rotation of the drum; and the thickness of sections corresponding to these intervals should be respectively .015, .0075, .005, .001^{mm}.

The New Object-Holder.—The object-holder which now accompanies the Thoma microtome is an invention of Mayer, Andres, and Giesbrecht. The object is now movable in all three directions. It is raised or lowered, and turned about the perpendicular axis by free-hand; but in the two other planes it is moved by pinions, so that the plane of section may be altered at pleasure during the process of cutting. As seen in Fig. 7, the object, o, imbedded in a small piece of paraffine, is attached to a larger mass of paraffine contained in a hollow metallic cylinder. The cylinder may be slipped up or down in the cubical block, A, and turned around its longer axis by means of a small metallic rod applied in holes near its lower end (holes not seen in figure). The position of the cylinder may be fixed by the handle, b, which works like the handle of a vice.

The block, A, may be turned about the transverse axis of the frame which holds it, by the pinion, D, and fixed by the screw, E, the head of which is provided with holes for the metallic rod.

In the same way, the frame itself may be made to turn about its longer axis by means of the pinion, G, and fixed by the screw, F.

The chief merit of this holder lies in the fact that the object

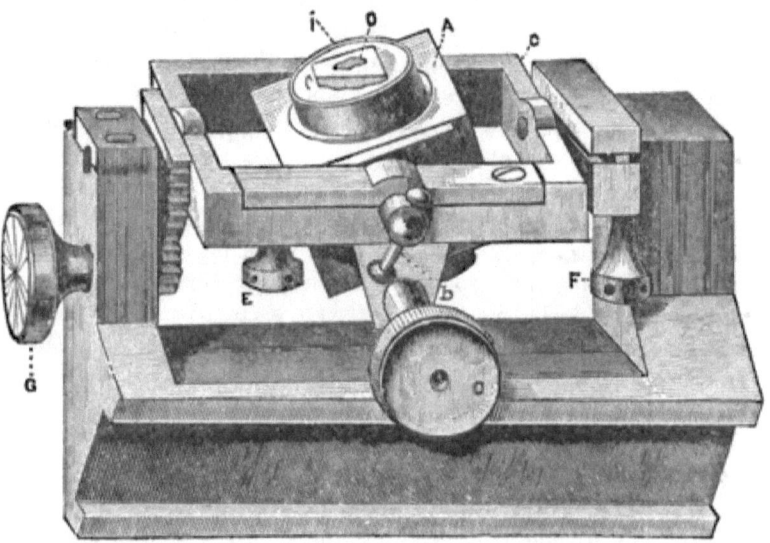

Fig. 7.—The Object-holder and Carrier.

may be rotated very freely about the two axes of the frame, without, at the same time, being raised or lowered very much. This latter advantage depends on the fact that the axes of rotation are near the top of the block, A; i. e., as near the object as possible.

The attachment of the object to a cylinder movable in a perpendicular direction has the great advantage that pieces of more than two centimeters in length may be sectioned. In order to obtain room for pieces of greater length, washers of

.5, or 1^{cm}, thickness may be placed at first under the knife, and afterwards removed.

An Improvement in the Carriers. — The advantages of having the carriers slide on five points (instead of two even surfaces) between two even planes have been thoroughly discussed by Thoma. The most recent improvement consists in making the so-called "points," of ivory, and the planes on which the points slide of an alloy of zinc and copper (Rothguss). The result is, that these parts are no longer exposed to rust, and that the plane surfaces on which the knife-carrier slides are not exposed to injurious friction.

Prices of the Microtome and its Adjuncts: —

The older microtome, consisting of a stand of cast-iron and the two carriers, without the micrometer-screw . . . . $24.00 (95 marks)
Same with ivory "points" and new object-holder . . . . 27.00 (110 ")
The registering micrometer-screw . . . . . 10.00 (40 ")
Knife for oblique cutting (16^{cm} long), with étui . 4.75 (19 ")
Knife for transverse cutting, with étui, holder, and clamp . . . . 4.75 (19 ")
Section-smoother . . . . 2.25 (9 ")
The complete set, including stand with "planes" of zinc and copper alloy, and ivory "points," registering micrometer-screw, the two kinds of knives with étuis, and section-smoother . . 53.00 (212 ")

This microtome can be obtained from the maker, Rudolph Jung, 15 Hauptstrasse, Heidelberg, or from the Educational Supply Company, 6 Hamilton Place, Boston, Mass.

THE SCHANZE MICROTOME.

This instrument will be easily understood from the figure. It combines both the screw and the sliding movement. The knife-carrier slides, as in the sledge microtome, but the object is raised by a screw, provided with a graduated drum, (Fig. 8, S). The Zeiss microtome, described by Körting,[1] is constructed on the same plan.

---

[1] Jen. Zeitschr. XIV., pp. 193–195, 1880.

The instrument is made of brass, or, preferably, of iron, with nickel-plated surfaces. The micrometer-screw moves the vertical plate, P, which slides up and down between two planes, and bears the object-holder. By means of screws, two of which are placed at right angles to each other, the holder can be inclined in any direction, raised or lowered, and rotated on its own axis.

FIG. 8.—Schanze Microtome.

One rotation of the drum, S, raises the object $.3^{mm}$; division of the drum into 30 degrees regulates the movement to hundredths of a millimeter.

CALDWELL'S AUTOMATIC MICROTOME.[1]

This machine has been devised to save labor to the histologist by cutting a very great number of sections suitable for microscopic investigation in a very short time. The machine is worked by hand, and may easily be made to deliver, in one continuous band, accurately cut sections at the rate of 100 per minute. To use it, however, to the best advantage, it is well to drive it by means of some motor, the fly-wheel being already

---

[1] *Quart. Jour. Micr. Sc.*, XXIV., Oct., 1884, p. 648.

provided with a groove for the reception of the cord coming from the motor. Where there is sufficient pressure and supply of water, a simple form of water motor seems the most appropriate and least expensive.

Method of using the Microtome. — Place one of the cylindrical vessels supplied with the machine upon a piece of paper on a glass plate, and pour into it sufficient melted paraffine to fill it. As this cools, the paraffine will contract, and will leave a hole, which must be filled up with more melted paraffine.

Melt a small quantity — say an ounce — of imbedding material in some suitable vessel; a small copper pan or a porcelain crucible answers very well, if care is taken not to allow it to become hotter than is sufficient to thoroughly melt it. Take a piece of glass, and smear it with a very small quantity of glycerine, to prevent the imbedding material from sticking to it. Then pour the melted material on the glass in small quantities at a time, so as to get a layer nearly a quarter of an inch thick. This, when cut up into suitable pieces with a knife, does very well for imbedding small objects. If larger objects are required, it is well to have two pieces of brass of the form shown in Fig. 9, 5, which, when placed together, will form a cavity half an inch in depth and of any desired length up to an inch or more. This cavity may be filled with the melted material in the manner already described, and the object to be cut must then be placed in position while the material is fluid. *It is well to cool the material as rapidly as possible by placing it in water as soon as it is sufficiently set.* From the cake thus formed, or from the piece cast in the mould, cut the piece of the material containing the object, and with an old scalpel, heated in a Bunsen flame, melt a small hole in the paraffine contained in the cylindrical vessel (Fig. 9, 1, a), and insert the piece of imbedding material containing the imbedded object; then with the heated scalpel melt a little of the paraffine round the base of the projecting piece, so as to give it firm support, and allow this to become thoroughly set.

Now remove the large brass plate from the top of the microtome (Fig. 9, 1, b), and insert the vessel containing the imbedded

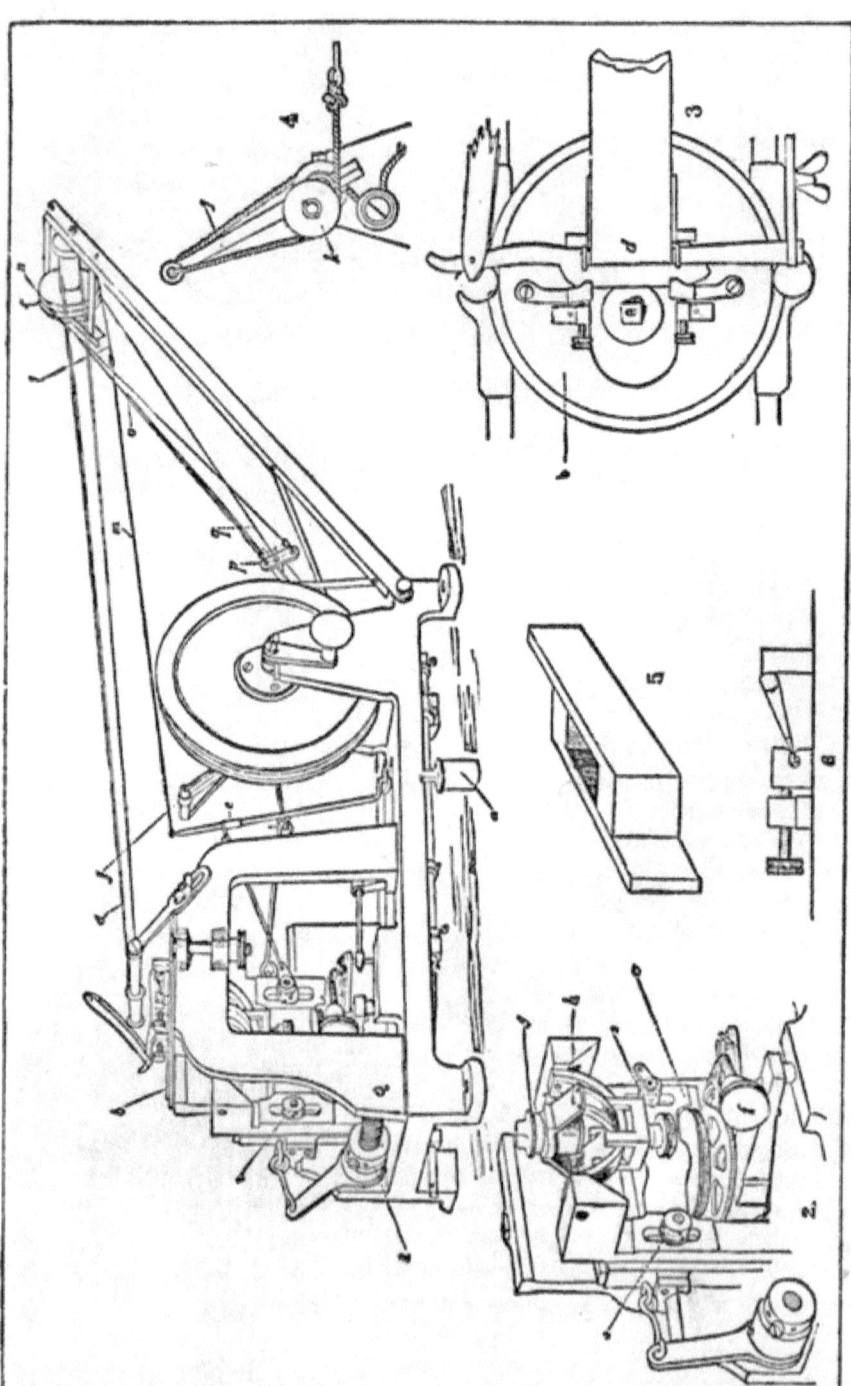

Fig. 9.—Caldwell's Microtome.

object in the tube for its reception, having first oiled the tube slightly to prevent the vessel from sticking. Next with a sharp knife *cut the material with the object imbedded in it so that all its opposite sides are parallel.* This is extremely important. Replace the top plate, and fix the razor in the holder provided for the purpose. The clamp is so made, that, if a little care is taken, the edge of the razor will not be injured (Fig. 9, 6). The razor must be set so that its back is as high as possible, as shown in Fig. 9, 6, and, above all, *the razor must be extremely sharp. It should be sharpened on a stone, and not on a strop.* The sharpness of the razor and the accurate parallelism of the sides of the mass to be cut are the most important points in the whole process.[1] Underneath the frame containing the object is a large brass milled head (Fig. 9, 2, c). By turning this the object may be raised or lowered according to the direction in which it is turned. This should be done until the object is just below the edge of the razor. The plate holding the razor should then be moved so that the edge of the razor is close to and quite parallel with the mass of material to be cut (Fig. 9, 3).[2] The plate should then be clamped by the screws at each side of it. A few turns of the fly-wheel will now bring the razor in contact with the object to be cut. The band of black ribbon (Figs. 9, 1 and 3 *d*) is now to be placed so that the end of it should be just above the razor, and clamped in that position.

---

[1] The makers of the instrument have nearly completed an automatic machine for sharpening razors, since it has occurred to them that this is an operation which may be performed with much greater accuracy by mechanical means than by hand.

[2] The distance through which the sliding stage moves can be altered by raising or lowering the arm (Figs. 9, 1, and 9, 4 *f*). This distance should be so arranged that the surface of the imbedding mass containing the object to be cut just clears the razor when the sliding carriage is at its maximum and minimum distance from either end of the machine. This is important, as the speed with which the black band travels varies directly with the throw of the machine. If this adjustment is made, and a little care is used in adjusting the ring (Fig. 9, 1, *e*), — see below, — the ribbon will move at each turn of the fly-wheel through a distance equal to the breadth of the surface which is being cut. If, on the other hand, the object swings far beyond the razor, the band will travel too quickly, and probably break the string of sections.

When the handle is turned, the sections should come off the razor in the form of a ribbon.

The ribbon of sections will not find its way to the continuous black band without assistance. With a needle in a handle, or with the point of a scalpel, pick up the end of the ribbon, when a sufficient length of it has been cut, and place it on the black continuous band, up which it will travel. When it reaches the top of the band, suitable lengths may be cut off with a pair of scissors. It may be found that the black band travels either too slowly or too fast. Its speed may be varied by moving the ring (Fig. 9, 1 e) up or down upon the vertical brass arm — upwards, if it is moving too fast; downwards, if too slow. A frequent cause of failure in the proper movement of the band is that the ebonite roller at the bottom of it is allowed to press against the razor; this must be avoided.

Varying the Thickness of the Sections. — In Fig. 9, 2, will be seen a milled head, f, which, when turned, controls the movement of the clicks, which, acting upon the ratchet-wheel attached to the micrometer-screw, regulate the thickness of the sections. This may be done so as to allow the clicks to engage one half, one, or several teeth of the ratchet-wheel, as may be required. When arranged for one half tooth, the sections will be $\frac{1}{10000}$ of an inch (.0025^{mm}) in thickness; when arranged to engage a whole tooth, $\frac{1}{5000}$ of an inch (.005^{mm}), and so on. At first it is well to use a whole tooth, as when thinner sections are cut so much depends on the sharpness of the razor. After cutting for some time, the machine will suddenly stop, the object ceasing to rise when the handle is turned. This means that the full extent of the micrometer-screw has been reached. It is necessary then to turn the large milled head (Fig. 9, 2 c) downwards, which will allow the carriage containing the object to fall to its lowest limit. It will be necessary now to raise the socket (Fig. 9, 2 g) in which the object is held, so as to be in position to come in contact with the razor. This milled head (c) is useful for rapidly getting the object in proper position, and avoiding considerable loss of time in turning the handle. The frame (h) which holds the socket is arranged with two quadrants, so that the socket may be set at any angle desired, and may be clamped with the milled head underneath it. This is for use when the object has not been symmetrically imbedded. The nut (Fig. 9, 1 *i*) is for tightening up the spring which draws the carriage of the machine back after having been pulled forward. In case this does not work properly, it is only necessary to unloose the two screws, and, with some strong but blunt pieces of steel placed in the two holes, to rotate the nut, so as to give a proper tension to the spiral spring. When this is done, the screws should be tightened up again to keep the nut in place.

The lock-nuts (Fig. 9, 1 and 2 *s*) should be screwed up sufficiently tight to barely prevent the carriage from falling by its own weight, so that, when the milled head (Fig. 9, 2 *c*) is screwed down, a slight pressure with the finger is necessary to make the carriage fall.

To arrange the machine for cutting different-sized blocks of material, it is only necessary to raise or lower the arm (Figs. 9, 1 and 4 *j*). When this arm is in a vertical position, the machine is arranged for its maximum traverse; when turned to the right and placed horizontally, it is at its minimum traverse. The cord, however, must always be in the groove of the wheel, *k*.

It is important to keep the strings which give motion to the endless band in proper position. The string (Fig. 9, 1 *l*) should go from the end of the wire, *m*, round the groove, *n*, in the pulley, and thence to the elastic band, *o*. The elastic band, *o*, should be stretched, and placed over the hook attached to the arm, *p*, care being taken that the shorter end of the arm, *p*, is uppermost. The string, *q*, should be tied to the stud upon which the arm, *p*, is supported, going thence round the groove, *r*, of the pulley, and back again to the hook at the longer and lower end of the arm, *p*, to which it should be tied.

Method of preparing the Slide. — Make by the aid of heat a viscid solution of white shellac in light-colored creosote. Spread a smooth, thin, and even layer of this solution on a clean, dry slide, with a camel-hair brush or with the little finger. Arrange the ribbon containing the sections on this slide while moist, and place it in the dry shelf of the water bath, which should be at a temperature slightly above the melting point of the imbedding material used. It should be left here until the creosote has evaporated, and the imbedding material melted. Now allow the slide to cool, and then wash it with turpentine until all the imbedding material is dissolved. Canada balsam in chloroform or turpentine and the cover-glass may now be applied in the usual manner. For convenience of mounting, it is extremely important that the ribbon of sections should be quite straight; and in order to insure this, it is necessary that the sides of the imbedding material from which the sections are cut should be quite parallel. The straight ribbon, when obtained, should be removed to some clean surface, and there cut into lengths appropriate to the size of the cover-glass used. It will be found convenient to use covers at least two inches long; indeed, a useful length for slides and cover-glass is six inches for the former and four inches for the latter.

CARE OF THE MICROTOME.

The planes on which the carriers slide should be cleaned and oiled freely every day when in use. The registering micrometer-screw is another part that requires frequent cleaning and oiling.

Neat's-foot, black-fish, or watch oil will serve the purpose. Ordinary sewing-machine oil is not to be recommended, as it gums, and thus greatly increases the wear of the instrument and the labor of cleaning. Minot recommends "pure paraffine oil, specific gravity 25." In drawing the knife, no pressure should be exerted, for this soon injures the knife and impairs the movement of the carrier.

RUTHERFORD'S FREEZING MICROTOME.[1]

This instrument (Figs. 10, 11) consists of a plate of gun-metal, B, with a circular opening at its centre. The opening leads into a well, A, closed inferiorly by a brass plug (K, Fig. 11), capable of being moved up or down by means of a screw, D. The tissue to be frozen is placed in the well, and sections are made by

---

[1] Outlines of Practical Histology, pp. 164-169, 1876.

gliding a knife through the tissue that projects above the level of the brass table. The thickness of the sections is regulated by an indicator. The freezing mixture is placed in the box, C, the water from which flows away by the tube, H. The microtome is clamped to a table by the screw, F. The object is seen in A, Fig. 11, and the freezing mixture in G of the same figure. The brass box is covered with gutta percha.

The knife should have a back with straight edges, so that it may not tilt in the process of sectioning. The surface that rests on the plate should always be hollow, and it is convenient

FIG. 10.— Rutherford's Freezing Microtome.

to have the upper surface the same, unless the sections be very large, when it should be flat or slightly convex.

The plug (K, Fig. 11) is unscrewed and oiled to prevent its becoming fixed during the freezing. The object and an imbedding mass are placed in the well.

A thick solution of gum answers well as an imbedding agent, as it does not become crystalline when frozen, and can be cut like cheese. If there be any alcohol in the tissue, it must be thoroughly removed by immersion in water from six to twenty-four hours, according to the size of the tissue, before it is placed in the gum. *It is always advantageous to allow the tissue to soak in the gum for twelve to twenty-four hours before freezing, in order that the gum may permeate every part of the tissue, and prevent the formation of a crystalline condition within the frozen tissue.*

Thick gum solution is poured into the well; and, when a film of ice has formed at the periphery, the tissue should be introduced, and held against the advancing ice until it is fixed. In this way the tissue may be secured in any position for the

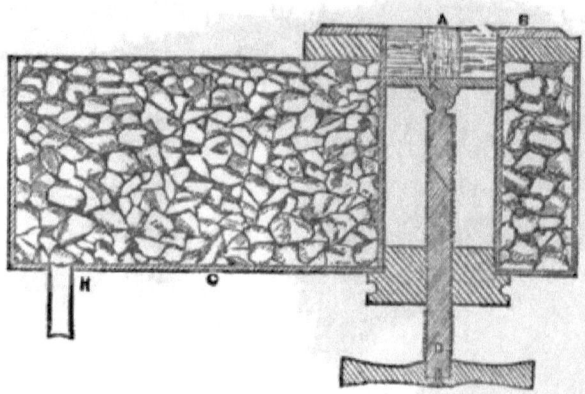

FIG. 11.—Rutherford's Freezing Microtome, sectional view (for explanation see text.)

process of sectioning. A plate of cork with a weight on it is placed on the microtome plate to prevent the entrance of heat, and also of salt from the freezing mixture.

Small quantities of *finely* powdered ice and salt are placed alternately, with the aid of a small spoon, in the freezing box, and are stirred round the well, the tube, H, being carefully kept open to allow the water to escape from the melting ice. When snow can be obtained, it is, of course, used instead of ice. The powdering of the ice is rather tedious, but it may be readily done in that form of sausage machine where two cylinders with spiral grooves are rolled against one another. Before putting the ice into this apparatus, it should be wrapped in a thick cloth and broken with a wooden mallet. The freezing process is really very simple, and can be fully carried out in from ten to twenty minutes. Of course the period is shorter, the colder the surrounding air.

It is not necessary to wet the knife, for this is thoroughly done by the thawing gum. The sections often contain a number of air bubbles; these, however, together with the gum, are soon removed by placing them in dilute alcohol (rectified or methylated spirit, 1; water, 2 parts). If the sections are to be

FIG. 12.— Jacobs' Freezing Microtome.

preserved for a time, until they are examined or mounted, they are transferred from the dilute alcohol to rectified spirit. When sections are apt to be spoiled by manipulation, they should be placed on the slide before they thaw. They can then be readily washed by making a pool of dilute alcohol around them, which may be changed two or three times.

A NEW FREEZING MICROTOME.

Dr. F. O. Jacobs has devised the freezing microtome seen in Figs. 12 and 13. It consists of a copper rod, A, two inches or more in diameter, and six inches high, enclosed by an inner zinc (b) and an outer brass tank (c). Above is the table, D, working on a fine screw (a). Through the centre of the table passes a narrower portion of the copper rod, the piston (p).

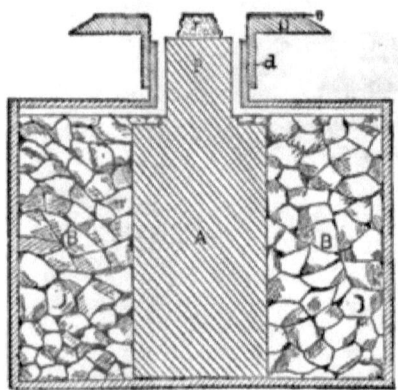

Fig. 13.—Sectional view of Jacobs' freezing microtome.

When the inner tank is filled with a mixture of salt, ice, and water, the temperature of the copper rod is so reduced as to freeze any object (F) placed on its upper end. The size of the rod is such that its temperature will remain very steady for from four to six hours without any further care on the part of the operator.

By this arrangement, — the advantages of which will be readily seen, — objects can be easily frozen, and without any slop or "muss."

The imbedding mass is composed of —

Gum arabic	5 parts
Gum tragacanth	1 "
Gelatine	1 "

The mixture is dissolved in enough warm water to give it the consistency of thin jelly when cold. A little glycerine (1 : 6) is added to the water.

AN ETHER FREEZING APPARATUS.[1]

A very simple and convenient little freezing apparatus, which can be used with almost any microtome, has recently been de-

---

[1] Zeitsch. f. Instrumentenkunde, Apr., 1884, pp. 126, 127.

scribed by W. Emil Boecker. Following the principle of the Lewes and the Roy instruments,[1] the ether spray is thrown upon the under side of the object plate, instead of on the object itself.

The apparatus consists of a short cylinder (Fig. 14, A), closed above by the plate, P, on which the object is placed for freezing. A metallic tube (a), drawn to a fine point, penetrates the base of the cylinder, and another (b) its side. The vertical tube (a) is connected by rubber tubing with a small bellows, while the horizontal tube (b) is similarly connected with a glass tube which passes through the stopper of a bottle containing ether, reaching nearly to the bottom. When the bellows is set in motion, the current of air throws the ether spray against the plate (P), and the rapid evaporation thus produced soon lowers the temperature sufficiently to freeze the object.

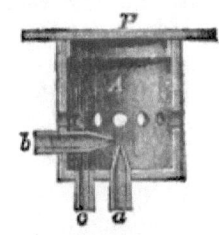

FIG. 14.— Boeker's freezing apparatus.

The ether vapor and the air escape through a series of holes in the side of the cylinder. A third tube (c) connects with a second glass tube that extends barely through the stopper of the bottle of ether, and thus provides for the admission of air into the bottle, and for the escape of the excess of condensed ether.

The cylinder is small enough to be received by the holder of the microtome.

It may be obtained from W. Emil Boecker, in Wetzlar, Germany, at sixteen and a half marks ($4.25).

The Roy microtome, referred to above, is made by Schanze Pathologisches Institut, Liebig-Strasse, Leipzig.

NOTES ON THE USE OF THE FREEZING MICROTOME.[2]

The tendency at the present time is to make all microscopic sections by the dry method after paraffin infiltration and imbedding; but no doubt there is a place, and an important one, for

---

[1] Arch. f. mikr. Anat., XIX., pp. 137-143, Pl. VI., 1880.

[2] From *Science Record*, No. 6, April, 1884.

the freezing microtome in practical histology, and in this note I would like to call attention to what seem to me improvements in its use. Disliking greatly the disagreeable "muss" made by ice and salt, it occurred to me to take advantage of the device of plumbers to thaw out water and gas pipes, — to use strong alcohol with the ice or snow instead of salt. By using snow, or finely powdered ice, and 95 % alcohol, a temperature of twenty centigrades below zero is obtained within five minutes, and this temperature may be maintained with far less trouble than with ice and salt. The microtome used is the Rutherford pattern, modified by placing the drain near the top instead of in the bottom. A rubber tube passing from this drain to a jar preserves the overflow. It requires about 250 cc. of alcohol to freeze and keep frozen one tissue for cutting; but this is not lost, as little evaporation takes place, and the dilution does no harm for many purposes; hence the method is not wasteful, while it is much more pleasant and expeditious than with salt.

Ordinarily tissues are infiltrated with thick gum before freezing, and then the sections are soaked in a relatively large amount of water, to remove the gum. Evidently, while soaking, staining, and transferring the sections, especially if they be of such an organ as the lungs, there is every liability of their becoming folded or torn. This may be avoided by staining the tissue in the mass, as for dry section-cutting, and then soaking in water to remove any alcohol, and finally completely infiltrating the tissue in a thick solution of very clean gum arabic. When ready to make the sections, the well of the microtome is filled with the thick gum, and the tissue introduced at the proper time, as usual. Before cutting, the gum is cut away from the tissue as in sharpening very bluntly a lead pencil; then, as the sections are cut, they are transferred directly to the slide. After several slides are filled, a drop of glycerine is added to each section, and the cover-glass applied. This is practically mounting in Farrant's solution. — *S. H. Gage.*

THE USE OF GELATINE JELLY IN THE FREEZING MICROTOME.[1]

Instead of freezing in gum, as is usual, one may use gelatine jelly. This is prepared and clarified in the ordinary manner. It should set into a stiff mass when cold; how stiff will best be learned by experience.

The tissue to be cut is transferred from water to the melted jelly, and should remain in it till well permeated.

It is then placed on the piston of a Rutherford's microtome; the "well" should not be filled; for adherence, it is sufficient to roughen the surface of the piston with a file. No more jelly should be used than is sufficient to surround the specimen; if too much has been added it may be removed, when frozen, by careful paring.

When well frozen, sections may be cut in the ordinary way; while frozen, they should be quickly transferred to the glass slide on which they are to be mounted. On touching the glass, the slice of jelly almost immediately thaws and adheres as a consistent fibre to the surface. When enough sections have been placed on the slide, they should each be covered with a drop of glycerine (the sooner this is added the better); a cover-glass is then superposed, zinc white or some similar cement is run round it, and the preparation is complete. In process of time the glycerine will permeate the gelatine and convert it into glycerine jelly; if this does not take place soon enough, it may be hastened by placing in an oven kept at a temperature of about 20 to 30° C.

In this way a series of entire sections may be obtained from the most disconnected structures, even when they contain hard silicious spicules, as in the case of sponges.

Diatoms may be cut without difficulty by this method.

MICROTOME KNIVES.

The Scientific Instrument Company of Cambridge, England, are preparing an automatic machine for sharpening microtome

---

[1] J. Sollas. Quart. Jour. Mic. Sc., XXIV. pp. 163-164. Jan. 1884.

knives, which, it is to be hoped, may prove to be just what every microtomist so much needs.

It is easy enough to find men skilled in honing ordinary razors, but not so with the microtome knives. Numerous and repeated trials have been made among the most hopeful experts in and about Boston, but in no case have the knives been returned in an acceptable condition; and sometimes they have been much damaged or wholly ruined. The best test for the condition of the edge is to try it on the palm of the hand. A knife that will not cut a ribbon of paraffine sections $.005^{mm}$ thick is not fit to use; the best knives should cut as thin as $.001^{mm}$. It is not often that it becomes desirable to cut so thin, but it is important in making thick sections ($.02-.01^{mm}$) to use a knife that has a much finer working capacity. A thoroughly sharp blade may have very nearly a horizontal position for its lower (plane) surface in sectioning, while a duller one requires to have its back raised a little above the level of its cutting edge. It is safe to say that a knife cuts well, when *thin sections* ($.005^{mm}$ or less) *agree in size with the cut surface of the paraffine block*. It may be possible to cut a straight ribbon with a dull knife, but in this case it will probably be found that the sections are shortened in a direction at right angles to the edge of the knife, which shows that the knife is acting the part of a plough, which crushes more than it cuts.

The statement that a sharp knife may have a nearly horizontal position must be understood to have some limitations. In general it may be said that the larger and harder the object, the more imperative it becomes to have the under surface of the blade slant towards the object; and the necessity for this is greater with a transverse than with an oblique knife. For very hard objects, a relatively thick-edged knife is required, as well as a slanting position.

For ordinary histological or embryological work, the upper surface of the blade is ground hollow, the lower surface plane (Fig. 14, *k*), the edge being left very thin, so that only an extremely slight bevel is made in setting. What bevel there is should be mainly on the upper side. The edge, when examined with a magnifying power of a hundred diameters, should be perfectly straight and *smooth*.

Method of Sharpening. Microtome knives can be properly sharpened only by those who understand their chief peculiarities, and who have trained themselves in this special work. The difficulties in acquiring the art are not, however, insurmountable; for, with the proper means and a little perseverance, they can be mastered in a short time. The first important step is to provide one's self either with a good razor-strop (those made by Zimmerman in Berlin are considered excellent), or with a long and wide oil-stone of the finest quality. Straps made of a leather band, unsupported by a solid base, and kept tense by the aid of a screw working in a frame, should never be employed in sharpening these knives; for they invariably give a bi-convex edge, with which it is impossible to do fine work. To secure a *plane* bevel of the cutting edge, the surfaces of the strop must be perfectly *smooth*, *flat*, and *hard*. In using the strop the knife is drawn back and forth, back foremost, *without pressure*, until the edge appears sharp when tested in the manner before mentioned.

FIG. 15.—Diagram illustrating the position of the knife in sharpening. *k*, knife; *s*, oilstone; *w*, wire.

In using an oil-stone, it is well to cover the surface of the stone with a mixture of glycerine (2 parts) and water (1 part), as recommended by Fol.[1] The blade is laid flat on the stone, and pushed *forward, edge foremost*, in such a manner that the free end of the knife finishes by resting on the more distant end of the stone. Here the blade is turned on its back and returned, edge in advance as before, to the place of starting. In drawing the blade, the utmost care should be taken *never to raise, in the slightest degree, the back from the stone;* and further, *the knife must not be pressed on the stone, but held lightly by the finger tips*, and the *necessary friction be left to capillary adhesion*.

After drawing the knife fifteen or twenty times, it should be tested as before.

---

[1] Lehrbuch d. vergl. mikr. Anat., p. 129. 1884.

The knives furnished with the Thoma microtome should be provided with a wire support (Fig. 15, *w*) for the back of the knife during the process of sharpening.

PREVENTING THE ROLLING OF SECTIONS.

The section-smoothers described below have been devised for the prevention of section-rolling.

Besides the section-smoother, there are other means by which the rolling of sections may be prevented. It may be effected by holding a thin narrow spatula over the edge of the knife while cutting. The spatula may be made of brass, in the form of Fig. 16 (*A*); or of paper fastened to a flattened needle, as indicated in *C*. The spatula should be bent slightly (*B*), and its convex face held over the paraffine without pressure. A small brush, slightly flattened, may be used for the same purpose.

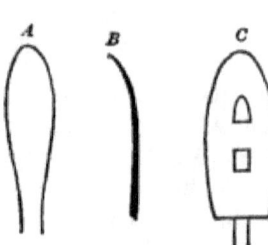

FIG. 16.—Section smoothers.

Rolling may also be prevented by making the paraffine softer and less elastic, through the addition of a small quantity of vaseline, and lastly — and most effectively — by placing the knife at right angles to the carrier.[1] The discovery of the fact that sections may be cut without rolling, by giving the knife a transverse instead of an oblique position, was made by Mr. Cadwell. Since the discovery of this method, it has come very rapidly into general use, and now Jung's microtome is supplied with 'transverse' as well as 'oblique' knives. This method, excellent as it is, especially with small objects, does not suffice in all cases, and does not therefore remove the necessity of a section-smoother. Even with the knife placed transversely, the section-smoother may often be used with advantage, and sometimes proves indispensable.

In this method it is important to use a moderately soft paraffine, which may be obtained by mixing, in proper proportions, soft and hard paraffine, and further, to give the piece of paraffine

---

[1] *Vide* Patten's method, p. 142-143.

to be cut a rectangular form. The piece must then be so placed in the holder, that the side next to the knife is exactly parallel with the cutting edge. Thus placed, every section lies flat on the blade. The second section pushes on the first, adhering to its adjoining side; the third pushes on the first two, adhering to the second. A whole ribbon of sections may be cut in this way in a few moments, without danger of losing their serial order. Thus three very important points are gained; the sections remain perfectly flat, the cutting may be as rapid as the hand can move, and the order of the sections is preserved without trouble to the manipulator. Care must be taken only that the opposite sides of the paraffine are parallel, otherwise the ribbon will curve to the right or left, and the arrangement of the sections on the slide be less easily accomplished.

The method used until quite recently at the Naples Zoological Station for preventing rolling consists in holding a needle, spatula, or brush lightly over the paraffine during the cutting of each section. This method has now been abandoned for the one described below.

The Section-Smoother[1] devised by Mayer, Andres, and Giesbrecht.—The section-smoother ("Schnittstrecker") is an instrument designed to prevent the rolling of sections; it is attached to the knife itself, and thus accomplishes its work unaided by the hand.

The accompanying figure represents a portion of the knife, k, with the section-smoother attached, and a block of paraffine, p. A section of the knife and rod, f', is also given, in order to show the position of the rod above the edge of the knife.

The most important part of the instrument is the cylindrical steel rod, f, which is supported in a position exactly parallel to, and close above, the cutting edge of the knife. In this position the rod compels the sections to pass between itself and the knife. The parallel position of the rod in the vertical plane is obtained by rotating it about the axis, x, which turns in the hole c or c'; the parallel position in the horizontal plane is

---

[1] "Neuerungen in der Schneidetechnik," in Mittheilungen aus der Zoolog. Station zu Neapel, Vol. IV., p. 429, 1883.

reached through the screw, a and a'; and the vertical distance of the rod from the edge of the knife is regulated by the screw, b. The entire apparatus is slipped on at the end of the blade, and held fast by two springs that press upon the under surface of the blade. The rod and its holder, which rotates on the axis, $d\,d$, can be lifted up from the edge of the knife by the aid of the handle, e, and turned back far enough to admit

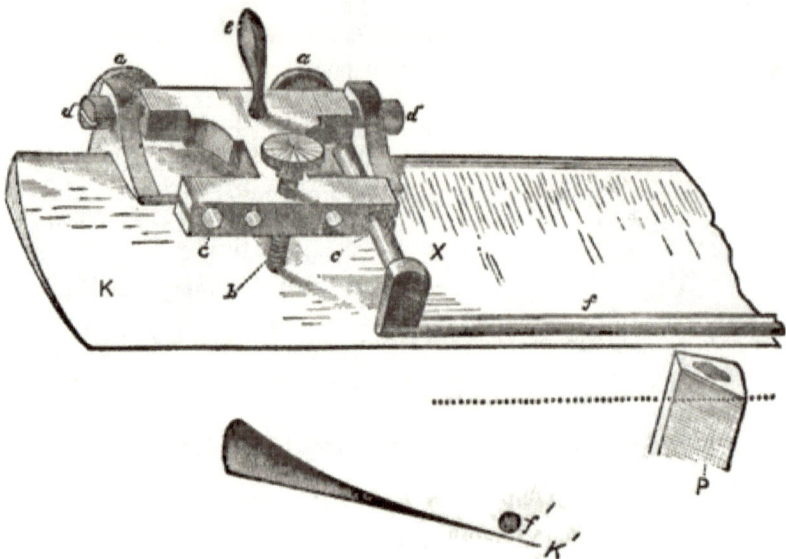

Fig. 17.—Section-Smoother. After Mayer, Andres, and Giesbrecht.

of cleaning the rod and the knife. The apparatus includes three interchangeable rods, differing in thickness in adaptation to sections of different sizes.

Schulze's Section-Smoother.[1]—This contrivance consists of a small weight supported by a steel spring. The weight (w), which has the form of a cylinder, is about 8^{mm} in length, and is fixed to the lower end of an upright rod, which can be turned about its longer axis, or moved up or down in a ferrule (f), as may be seen from the figure.

[1] Franz Eilhard Schulze. Ein Schnittstrecker. *Zool. Anzeiger*, VI., No. 132, p. 100.

One end of the spring (s) supporting the weight is soldered to the ferrule; the other end is held fast in the holder (k), one arm of which is prolonged into an enlarged handle-like portion. The holder (k) fits into a cylindrical hole, which penetrates perpendicularly to a depth of 2–3^{mm}, the hinder portion of the object-carrier. Within this hole the holder can be turned, or moved up or down; and these movements, together with those which may be given to the rod (r) and the spring (s), are ample for all adjustments of the weight (w).

Preparatory to cutting, the weight should be turned so as to be parallel with the edge of the knife, and raised or lowered by means of the rod (r) and the holder (k) until it rests *lightly* on the upper surface of the paraffine. Slight changes in the pres-

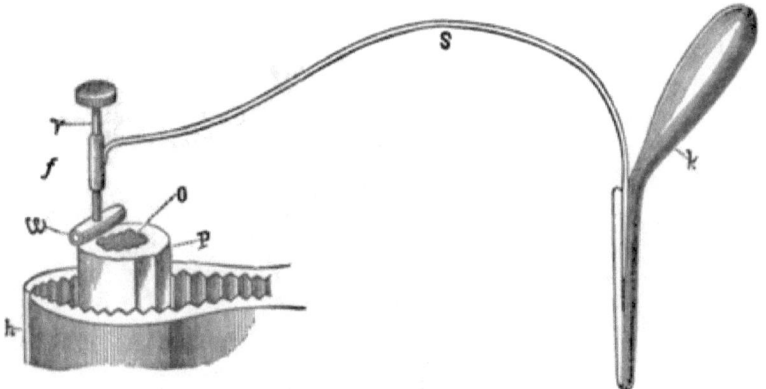

FIG. 18. — Schulze's Section-Smoother adjusted to a block of paraffine (p) preparatory to cutting; o, the object; h, paraffine-holder; s, steel spring; w, weight.

sure of the weight can be made through the rod (r); greater changes can be effected by bending the spring.

The weight should rest, not directly over the object, but on the edge of the paraffine next to the knife.

This section-smoother, which can be fitted to any sliding microtome, can be obtained of Fr. Fasching, in Graz, Bürgergasse, No. 13, at a price of 3½ marks, or 85 cents.

Decker's Section-Flattener. — Dr. F. Decker[1] has invented a section-flattener which presents an important improvement on those described above. The flattener is a cylindrical glass roller (a), which rotates on a brass rod (b)

The diameter of the rod is 2^{mm}, slightly less than the lumen of the roller, so that the latter rotates easily and evenly.

A steel plate (e), bent so that the two arms clasp the blade firmly, forms a carrier for the roller and its shaft. The carrier is slipped on at the free end of the blade, and the pressure of its arms strengthened by means of the screw (f), the inner end of which bears against a narrow strip of steel (g). To the upper arm of the carrier a brass block (h) is attached by means of a hinge, which runs lengthwise along the middle of the under

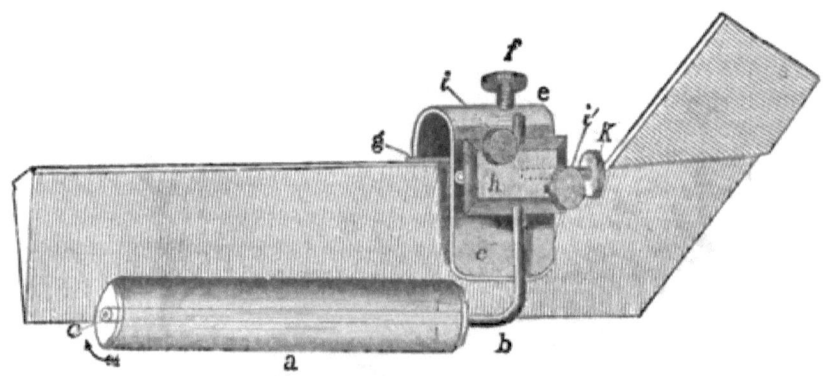

FIG. 19. — Decker's Section-Flattener.

side of the block. The short arm of the brass rod (b) passes through the middle of the block in a horizontal direction, and is fixed in any position by means of the screw (k). The distance of the glass roller from the edge of the knife is regulated by the screws, i and i'. If one of these screws is raised, and the other lowered, one half of the block (h) will be raised, and the other half lowered, and the distance between the roller and the blade correspondingly altered. The length of the glass

---

[1] Ein neuer Schnittstrecker. *Arch. f. mikr. Anat.*, XXIII., pp. 537-543, 1884.

cylinder (*a*) is 5^{cm}, its diameter should vary according to the size of the object to be sectioned. Three cylinders, measuring 4, 6, and 9^{mm} in diameter, will be found sufficient for all ordinary purposes.

The apparatus above described can be adjusted to knives of different microtomes. It may be obtained from Gustav Stöber, in Würzburg.

Kingsley's Section-Smoother.[1] — Led by a suggestion of P. Francotte,[2] Mr. Kingsley devised a simple form of section-smoother, which can readily be made by anyone, and which fairly answers the purpose for which it was intended. A piece of iron wire is bent in the manner shown in the cut, so that the two ends will form a spring clip, grasping the back of the knife. The middle portion is so fixed that it will be parallel with the edge of the knife, and at a distance of about a hundredth of an inch from it. For this purpose the writer has found an ordinary hairpin, deprived of its lacquer, about the right size.

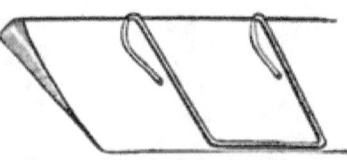

FIG. 20.— Kingsley's Section-Smoother.

In cutting, the section passes between the wire and the blade, and all tendency to curl is prevented. This form of smoother is applicable to the knife when used in any form of sledge microtome, or when cutting free-hand; but for use with the Sterling (well) microtome, it is evidently ill adapted, for the ends which come underneath the blade would interfere with the work. The form of smoother described by Drs. Gage and Smith embraces the same principle, but is more complicated.

The Section-Flattener of Gage and Smith.[3] — The section-flattener consists of a rod (*b*) of spring brass, about 5^{mm} in diameter, flattened on two sides (*B* and *D*), extending parallel with the edge of the knife, and projecting about 2^{mm} beyond it. Opposite the cutting edge, the space between the

---

[1] J. S. Kingsley, Science Record, 11, No. 5, p. 112, March 15, 1884.

[2] Bull. Soc. Belge de Microsc., X., 1883–4, No. 3, p. 55.

[3] "Section-Flattener for Dry Section-Cutting." *The Microscope*, February, 1884.

rod and knife is about 1^{mm}, while nearer the back of the knife the distance is greater (Fig. 21 *D*, *a*, *b*). At each end, the rod is bent at right angles. Next the handle, it passes through a hollow cylinder (Fig. 21, *d*), into which it is secured by a milled nut (*i*). At the free end of the knife, the rod is hooked over the back of the blade (Fig. 21, *A*), the spring of the wire securing it firmly. At the two angles of the rod it rests on the blade, so that, in cutting sections, any amount of pressure may

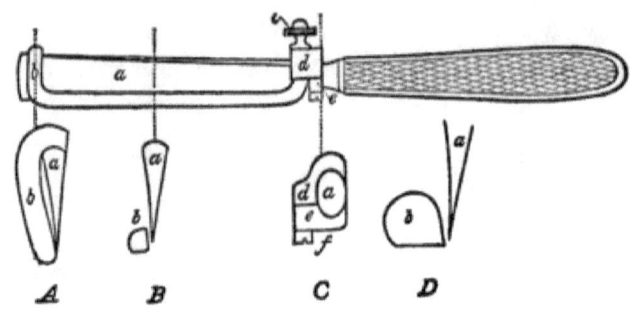

FIG. 21.—The section-flattener attached to a knife. (*a*), blade of knife; (*a'*) tang; (*b*), section-flattener; (*c*), milled nut; (*d*), the part of the clamp bearing the hollow cylinder; (*e*), part of clamp; (*f*), screw holding the two parts of the clamp together.
A, section showing the end of the flattener hooked over the blade; *B* and *D*, sections showing the form of the rod (*b*), and its position with relation to the edge of the knife; *C*, section of the tang, showing the manner of attaching the clamp.

be applied at these points. The rod is attached to the knife by means of a clamp, which consists of two pieces clasping the tang and held together by a screw (Fig. 21, *C*). To clean the knife and rod, or to remove sections, the rod may be raised, as it swings freely in the hollow cylinder attached to (*d*). The rod may be removed by removing the milled nut (*i*); the entire apparatus may be removed from the knife by loosening the screw (*f*).

CHAPTER IV.

METHODS OF IMBEDDING.

In every method of imbedding, the principle is the same; namely, to saturate objects with substances which not only fill the larger internal cavities, but which also penetrate the tissues themselves, rendering them (after cooling) sufficiently hard for the process of sectioning.

PARAFFINE.

One of the most convenient and most commonly used substances for imbedding is paraffine. The process of imbedding is, in general, very simple; and the mass is ready for cutting a few minutes after the operation is completed. Where fineness and uniformity of section and rapid execution are of importance, the paraffine method has no equal.

In this method, the *hardness* of the paraffine is an important consideration. There is no invariable rule to be followed in this respect; it is necessary to use different degrees of hardness, according to the *thickness of section* required, the *nature of the object*, and the *temperature of the room* in which the object is sectioned. It is convenient to have in stock a *soft* (fusible at 45° to 46° C.) and a *hard* paraffine (fusible at 55° to 56° C.), and to mix these in proportions suited to the occasion. As a general rule, for a room temperature of 68° to 70° F. (20° to 22° C.), it suffices to mix two parts of the hard with one of the soft paraffine. The harder the paraffine, the easier it is to obtain very thin sections, and the greater the liability to roll. The melting point of the different sorts of paraffine ranges between 40° and 60° C. The most general rule for determining the proportions of soft and hard paraffine is to so mix that the point of fusion lies between 50° and 53° C.

Dr. Minot employs paraffine melting at 55° to 56° C. and a soft "chewing-gum" paraffine melting at 50° C., and mixes in the proportion of 20 : 9.

For cutting ribbons of sections, Adam Sedgwick uses hard paraffine coated with soft. The block is coated by dipping it in melted soft paraffine, and the soft paraffine is then cut away from the two sides of the block, being left on the front and hind faces only.

Preparation for Imbedding. — Preparatory to imbedding, the objects are removed from absolute alcohol (96 % will be sufficient) to clove oil, turpentine, creosote, or chloroform, and left until thoroughly saturated. The penetration of the clarifying medium may, in some cases, be advantageously hastened by warming a little. In transferring from alcohol to the clarifying medium, there is often considerable danger of collapse and distortion from shrinkage, especially with delicate objects, and with such as have larger or smaller internal cavities, or are enveloped by impervious membranes. In the use of clove oil, which is one of the most satisfactory media, this danger may be guarded against by placing the object first in a mixture of alcohol and the oil in equal parts, and then, after some time, into pure clove oil.[1]

In order to avoid shrinkage, in passing delicate objects from alcohol to chloroform or an oil, Giesbrecht[2] recommends the following method: —

Pour a little absolute alcohol into a small glass tube, place the canular end of a pipette, containing the solvent, below the surface of the alcohol, and allow a few drops to flow from it to the bottom of the tube; into this tube let fall, by the aid of another pipette, or a small spatula, a few drops of absolute alco-

---

[1] As clove oil mixes very slowly with paraffine, it is sometimes advisable to place the object, after clarification in clove oil, in turpentine for a while, transferring from this to paraffine.

[2] Giesbrecht. "Zur Schneide-Technik," in *Zoolog. Anzeiger*, 1881, No. 92.

hol, containing the objects to be imbedded. The objects will sink through the alcohol, which, being the lighter fluid, has taken a superjacent position, and rest on the upper surface of the fluid expelled from the first pipette. Most of the alcohol may now be removed by a pipette, and the objects left to sink gradually into the heavier fluid at the bottom of the tube. In this way, the replacement of the alcohol contained in the objects by an oil, or some solvent of paraffine, is much retarded, and thus the danger from shrinkage reduced to a minimum.

Where chloroform is preferred to creosote or oil of cloves, a little ether (æther sulfuricus, C_4H_5O) should be added, as many objects will not sink in pure chloroform.

Dr. Giesbrecht recommends for difficult cases chloroform,[1] as it is one of the best, and, at the same time, the most volatile, solvent of paraffine.

After the objects have become thoroughly saturated with chloroform, the containing tube is placed on a water-bath, and heated to about 55° C., the melting point of paraffine; then a small piece of paraffine is added, and allowed to dissolve; and this is repeated until bubbles cease to rise from the objects. To make sure that the chloroform has been fully expelled, the objects may next be transferred to pure paraffine, and left for a few minutes before imbedding.

---

[1] Bütschli (*Biolog. Centralblatt*, B. 1, p. 591) has also recommended chloroform.

For the Hydrozoa, Professor Weismann prefers turpentine to chloroform, as, where the latter has been used, the paraffine is liable to be more or less spongy, in consequence of bubbles lodged in the tissues.

Turpentine renders objects brittle; and, on this account, chloroform will, in many cases, give better results. The spongy state of the paraffine results from the fact that the chloroform has not been allowed to wholly escape.

In the case of the Actiniæ, Dr. Andres employs a mixture of turpentine, creosote, and alcohol, using successively mixtures containing more turpentine and less alcohol, thus:—

	Mixture No. 1.	No. 2.	No. 3.	No. 4.
Turpentine	1	2½	4½	7½ parts
Creosote	2	2½	2½	2½ "
Alcohol (absolute)	7	6	4	

Kossman[1] says that chloroform is the only solvent that can be successfully used in the case of objects with thick chitinous membranes.

In the use of chloroform, two points must be carefully attended to, namely, the *complete saturation* of the object before it is placed in paraffine, and the *complete evaporation* of the chloroform before the object is finally imbedded. If the first point is not secured, of course the paraffine will not penetrate the object thoroughly; and if the chloroform does not wholly escape before the process of imbedding begins, the paraffine will be spongy, and consequently unfit for section-cutting. The evaporation of the chloroform may be effected in two or more ways. In all cases the object must lie in chloroform until thoroughly saturated. Then paraffine may be added gradually, as recommended by Dr. Giesbrecht; or the saturated object may be placed in a solution of paraffine in chloroform, as recommended by Professor Bütschli. After remaining here until it is thoroughly impregnated (an hour or less), it may be placed in a watch-glass with a little of the solution, and kept at a temperature of about 50° C. until the chloroform has escaped. In case of larger objects, they may be transferred directly from the solution to pure paraffine, without undergoing the slow process of evaporation.

Kossmann transfers the object directly from pure chloroform to pure paraffine, and allows it to remain in the paraffine (kept at a constant temperature of 50° C.) for several hours,— sometimes for two or three days.

Saturation with Paraffine.— After the object has been properly clarified by one of the above methods, it is to be transferred to melted paraffine. It should be soaked first of all in very soft paraffine, or a mixture of paraffine and turpentine, and then in the paraffine intended for imbedding. For this purpose a convenient form of water-bath is required, such as that described below, in which the two paraffines can be kept in separate basins, at a temperature of about 55° C. The time required for complete saturation with paraffine will vary ac-

---

[1] Zool. Anzeiger, No. 129, p. 20.

cording to the size and nature of the object, from 15 minutes to as many hours. For objects not more than one or two millimeters in thickness, 30 or 40 minutes, divided equally between the two basins, are usually sufficient. In the case of large objects, and such as are penetrated with difficulty, owing to membranes or other impediments, the process may be made much more expeditiously, and with far greater safety in vacuo, by the aid of Hoffman's imbedding apparatus.

Imbedding Boxes. — The saturation with paraffine being complete, the concluding steps in the process of imbedding may be undertaken.

Boxes for imbedding may be made of rectangular pieces of paper in the following manner: The paper is first broken in the lines $a\ a'$ and $b\ b'$, then $c\ c'$ and $d\ d'$ (by bending always towards the same side). Then in every corner a break ($A\ A'$, $B\ B'$, $C\ C'$, $D\ D'$), is made by bringing $A\ c$ and $A\ a$ together. The four sides of the box are next bent up, and the corners at the same time turned outward and back behind the ends $A\ B\ a\ b$ and $C\ D\ a'\ b'$. Finally the upper edge of these ends is bent down over the corners.

Bubbles around the object may be removed by means of a heated wire. (*Blochmann.*)

A more convenient box, introduced by Dimmock, may be made of two pieces of type metal (or better of brass). As will be seen from the accompanying diagram,

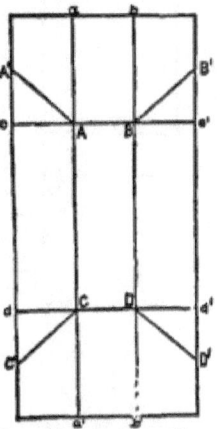

FIG. 22. — Diagram illustrating method of making paper boxes for imbedding.

each piece of metal has the form of a carpenter's square, with the end of the shorter arm triangularly enlarged outward. A convenient size will be found in pieces measuring 7^{mm} (long arm) by 3^{cm} (short arm), and 7^{mm} high. With such pieces a box may be constructed at any moment by simply placing them together on a plate of glass, which has previously been wet with glycerine, and gently warmed. The area of the box will evidently vary according to the position given to the pieces, but the height can be varied only by using different sets of pieces. In such a box the paraffine may be kept in a liquid state by warming now and then over a spirit lamp, and small objects be placed in any desired position under the microscope.

It is well to imbed in a thin layer of paraffine, so that the object, after cooling, may be cut out in small cubical or pyramidal blocks, which may be easily fixed, for cutting, to a larger block of hard paraffine, or, better, to a block of wood saturated with paraffine. (*Mayer.*)

Mr. Dimmock also used for the same purpose 'quotations', such as are used by printers in filling blank spaces at the beginning and end of chapters. These quotations vary somewhat in size, and are sold at 25 cents a pound.

Mayer, Andres, and Giesbrecht recommend brass instead of type-metal for these boxes, and a wash of thin collodion, where it is desirable to keep the paraffine in a melted condition for a considerable time, as in imbedding small objects in definite positions. The glass plate forming the base of the box is first wet with glycerine, and then the box is washed with collodion, and placed on a water-bath, in order to evaporate the ether. In this way a box is obtained in which paraffine can be kept for hours in a fluid condition, without escaping between the glass and the metallic pieces. The box is kept on a small water-bath, made for this special purpose, while arranging the object under the dissecting microscope.

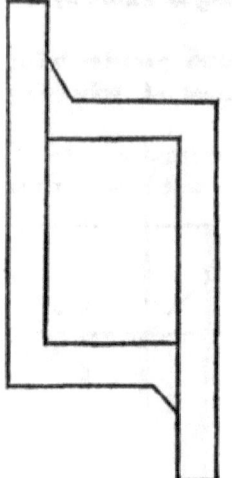

FIG. 22.—Type-metal box for imbedding.

Orientation. — With larger objects there are no difficulties in the next step. It is only necessary to fill the box with melted paraffine, and then to drop in the object by the aid of a spatula, and place it in any desired position with a hot needle.

Orientation becomes difficult only with objects so small that their position cannot be controlled except by the aid of a dissecting microscope. Round objects, less than one millimeter in diameter, e. g., many ova and embryos, — are the most difficult to manage. Such objects may usually be successfully oriented in the following manner: —

1. Prepare the box; for this it will be necessary to use the two triangular pieces of metal, a rectangular glass plate (2 in. × 2½ in.). The plate should be cleaned and then smeared with glycerine, and the pieces of metal so adjusted that the arms are parallel with the edges of the plate.

2. Having warmed the box over a spirit lamp, lift the object from the basin of paraffine by the aid of a *small, flat, thin* spatula (first starting it from the bottom by shaking the paraffine a little), and allow it to *flow* with the paraffine carried on the spatula into the box.

3. Then fill the box (5–6^{mm} deep) with the melted paraffine, and warm it a little over a spirit lamp, just enough to keep all of the paraffine in a liquid condition for a few moments. Now place the box on the warm table of a dissecting microscope, and by the aid of a hot needle proceed to place the object in the desired position. As the object is illuminated from below, it can be easily seen, turned over, and moved about at pleasure. If the paraffine freezes before orientation is effected, it should be melted again as before, and the needle must be kept hot by repeatedly holding it in the flame of the lamp.

The difficulty of finding very small objects in a basin of paraffine will be very much lessened by keeping the paraffine free from dust, and the bottom of the basin (tin) scoured bright. A piece of emery cloth serves for polishing.

The necessity of rewarming the box of paraffine, which often arises in the above method, may be removed by using a hot bath on the table of the microscope. This bath should be a box of convenient size (not over 2^{cm} high), with top and bottom of glass, with an opening at one end for filling with hot water, and another at the opposite end provided with a rubber tube and clamp, for drawing off the water as soon as the object has been arranged.[1]

Prevention of Bubbles. — After the imbedding process

---

[1] A similar device has been described by Fol, Lehrb. d. vergl. mik. Anat., p. 122.

has been carried thus far, there is still another danger to be carefully guarded against. If the box is left to cool slowly in the air, bubbles are very likely to appear in the paraffine, which will prove a serious obstacle in cutting. Profiting by Caldwell's suggestion, to cool the box in water, one may avoid all such inconveniences. As soon as the paraffine cools around the object, so that its position is secured, the box should be held in a vessel of cold water, first at the surface (until the paraffine has set), then fully submerged. In this way the paraffine is quickly cooled sufficiently for removal from the box, which may then be used for imbedding a second object. A dozen objects may be thus imbedded in a very short time. If the box is plunged below the surface of the water before the paraffine has become rigid, holes will arise in the mass and fill with water.

Dr. Schulgin's Mixture for Imbedding.[1] — Instead of pure paraffine, Dr. Schulgin uses a mixture of paraffine with ceresin, a substance somewhat similar to wax, but firmer and much less brittle. Paraffine, which melts at 55° C., is recommended; and the amount of ceresin to be added to a given amount of paraffine may be determined by experiment. The finest sections of this substance are not brittle, and herein lies the chief excellence of the mixture. If this mixture proves too hard, it may be softened by adding a little vaseline.

Caldwell's Method of Imbedding in Paraffine.[2] — After the specimen has been stained it should be left in 90 % alcohol for a few minutes, and thence transferred to absolute alcohol, there to remain until all the water is extracted. The length of time necessary for this varies greatly with the size of the specimen. A three-day chick, for instance, will require about an hour, larger specimens a day or more, in which case the absolute alcohol should be changed occasionally. Some tissues may be transferred directly from the absolute alcohol to turpentine, and thence in about two hours to the melted imbedding material. For delicate tissues, however, the

---

[1] Zool. Anzeiger, VI. No. 129, p. 21, 1883.
[2] Quart. Jour. Micr. Sc. XXIV., Oct. 1884.

following process, though longer and more troublesome, is preferable. With a pipette introduce some chloroform, to which two or three drops of ether have been added, under the alcohol in which the object is lying. The object will then float for some time at the junction of the alcohol and chloroform, and will finally sink into the chloroform when saturated with it. If, as often happens in the case of embryonic tissues, the object is lighter than the chloroform, it is not easy to tell when the saturation is complete, but generally, on shaking the bottle, a saturated tissue can be temporarily covered by the chloroform, while tissues containing alcohol keep steadily on the surface.

When the tissue is saturated with the etherized chloroform, it should be transferred to pure chloroform, and there left for a few minutes. Then drop in some pellets of soft paraffine and leave it for two hours or more, shaking occasionally. The whole should then be poured into a small melting-pot, and a quantity of imbedding material added. The melting-pot should then be placed in the water-bath at a temperature of about 60° C., and there left until all the chloroform has evaporated, which may be determined by the absence of smell of chloroform on shaking. If much imbedding material is required, this process takes a day or two; it is therefore better, when the solution of imbedding material is fairly strong, to take out the tissue and put it direct into pure melted imbedding material. In any case no chloroform must remain in the material to be cut, as it makes it brittle. Generally speaking, the more gradually these processes are passed through, the better will be the result.

Ryder's Method of Imbedding Fish Eggs[1] — which have been colored *in toto* with borax carmine, or borax picro-carmine, is as follows: —

a. After dehydration with about forty times their own volume of strong commercial or 97 % alcohol, and afterwards saturated with oil of cloves, the embryos are placed in a watch-glass containing a melted mixture of chloroform and pa-

---

[1] On the Preservation of Embryonic Materials, etc., p. 15. 1884.

raffine in equal parts, in which they may remain twenty or thirty minutes, at a temperature not above 150° F. When saturation is complete, the eggs have the same appearance in the melted mixture as in alcohol.

b. From the above they are transferred to another larger dish, containing pure paraffine, which melts at 158° F., but which must on no account be allowed to boil. Here they remain for twenty to thirty minutes more.

c. The embryos are then transferred, one or two at a time, to a common slide, such as is used for mounting objects. The slide may be warmed over an alcohol lamp. A brass ring, 5 to 8 millimeters deep, and 24 in diameter, is then placed on the slide around the object. This ring is then filled with melted paraffine, and the object arranged in it in the desired position, with a hot needle, when the whole is left to cool.

d. After cooling, the paraffine contracts within the ring, when the latter may be removed, and the discoidal block may then easily be loosened from the slide. The block may then be trimmed down, with a scalpel, into a shape suitable for fastening into the well in the carriage of a sledge microtome, or the block may be marked and laid away until it is wanted for use.

IMBEDDING APPARATUS.

In imbedding relatively large anatomical preparations, the penetration of the paraffine to all parts can be secured by the aid of a good air-pump; but this method is tedious, and robs one of time. The same end may be reached with very little cost of time and labor by means of a suction-pump, such as is used in chemical laboratories, provided only that the water pressure at command is sufficiently great to do the work. Dr. F. W. Hoffman [1] has described a simple apparatus to be used with the suction-pump.

The suction-pump, (Fig. 24 *S*), is connected with the exsiccator, *E*, by means of rubber tubing, through each portion of which runs a piece of glass tubing as a protection against undue com-

---

[1] Zool. Anz., VII., No. 165, p. 230, 1884.

pression. The exsiccator, holding the small basin of paraffine, *P*, in which is placed the thermometer, (*Th'*), stands in a zinc pan (*W*), containing water and a thermometer (*Th*). The capacity of the pan should be such that the temperature of the water can be kept quite even during the process of imbedding. The flask, *F*, introduced between *S* and *E* by means of a T-shaped tube serves to catch any water which may find its way into the connecting tube, in consequence of variation in the water pressure. The glass tube, *m*, the lower end of which is in a bottle of mer-

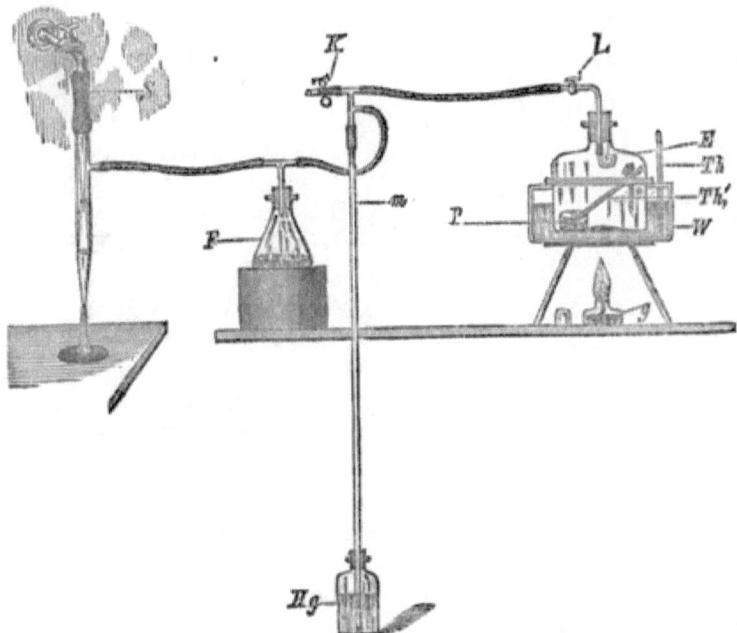

FIG. 24.—Hoffman's Imbedding Apparatus.

cury, serves as a manometer. The pressure indicated enables one to judge whether the preparation is saturated with paraffine.

In using this apparatus, the water-bath, *W*, should first be heated to a temperature of about 60°; then the basin, *P*, containing *melted* paraffine and the object to be imbedded, may be placed in the exsiccator and the pump set in operation. When the mercury reaches the highest point in the tube, m, and air-bubbles cease to rise from the object, the process is ended, and the air admitted by loosening the screw-clamp, K. Before admitting the air, the stop-cock, L, can be closed, in case it is thought best to leave the preparation still longer in vacuo. As soon as sufficient time has been given for complete saturation, the stop-cock, L, is opened slowly, and the air streams in. The end of the connecting tube is bent upward in E, in order that the paraffine may not be disturbed by the inflowing air. Finally, the object is taken out and placed in a box of melted paraffine, and left to cool.

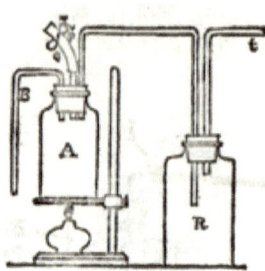

FIG. 25.—Francotte's imbedding apparatus.

With a water pressure of 700–720^{mm}, Hg., most objects will be completely saturated within 20 minutes.

If water pressure is wanting, a vacuum may easily be produced with the apparatus devised by G. P. Francotte.[1] This apparatus (Fig. 25) is connected with the exsiccator (E, Fig. 24) by means of the tube, t', and is provided with a barometer tube, B, which descends into a vessel of mercury. The vessel, A, holding about half a litre, is partially filled with water. The water is boiled by the aid of a spirit-lamp, and the air escapes by the tube, c, which is provided with a pinch-cock, P. When the steam passes out in a jet, P is closed, the lamp removed, and A cooled slowly by a wet sponge. The vacuum thus produced causes the mercury to rise above 70^{cm}. After half an hour, the air may be admitted. The flask, R, serves to receive the steam when formed in excess.

WATER-BATH.

A convenient form of water-bath, devised by Dr. Mayer, is represented in Fig. 26.

---

[1] Bull. Soc. Belg. Micr. XI., pp. 45–48, 1884.

It is a small brass box 18cm long, 9cm wide, and 8cm high. The tube, *A*, through which the water is received, and the rod, *b*, serve as handles. The receiving tube is closed by a cork, provided with a glass tube for the escape of steam, which is bent in the form of a siphon to protect against dust. One and a half centimeters from the base of the box is an oven, *o*, .7cm high and 12cm long, which passes completely through the box, and serves for warming the slides when shellac is used. Above are seen two circular, basin-like pits, P, 5.5cm in diam.

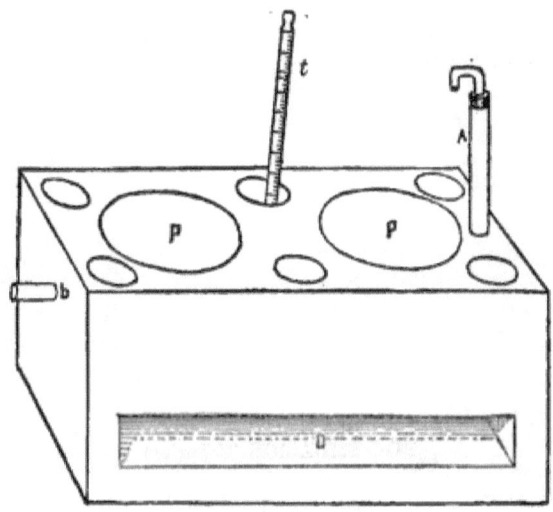

FIG. 26. — Water-bath.

and 4cm deep, for receiving the two tin paraffine cups. These are covered by circular plates of glass. There are also six tubular pits, one for a thermometer *t*, the others for glass tubes.

This water-bath will be found useful for other purposes than those of imbedding and mounting. It will, of course, be understood that the purpose in giving its exact dimensions is simply to furnish a guide where one is required. There are at least two important advantages offered by this water-bath over those in general use; viz., the slides are protected from dust, and the paraffine is not exposed to the water.

GUM ARABIC.[1]

The object, saturated with water,[2] is placed in a watch-glass, and surrounded with finely powdered gum arabic. The gum absorbs the water, and forms a mucilage which gradually penetrates the object. The preparation is next placed in a small cavity made in elder pith, and covered with the pulverized gum. The whole is left to dry. A little before complete dessication, it is plunged into 80 % alcohol. After remaining over night in alcohol, the solidified gum has acquired the consistency necessary for cutting. While cutting, the knife should be kept wet with 80% alcohol.

GLYCERINE AND GELATINE.

Gelatine	1 part
Distilled water	6 "
Glycerine	7 "

For preservation, a little carbolic acid (1 gram for 100 grams of the mixture) should be added. Objects are transferred directly from water to the melted mixture, and, after complete saturation, imbedded in paper boxes. After cooling, the objects thus imbedded are hardened in alcohol, then sectioned, and mounted in glycerine.[3]

EGG-EMULSION (Calberla).[4]

The white of several eggs is separated from the yolk, freed from the chalazæ, cut with shears, and thoroughly mixed by shaking with a 10 % solution of carbonate of sodium (fifteen parts of the white to one part of the solution). The yolk is next added, and the mixture shaken vigorously. After removing the foam and floating pieces of yolk by the aid of filtering paper, the so-called "egg mass" is ready for use. It is this

---

[1] Gravis, Bull. de la Soc. Belg. de Mic. No. VII., 1884, p. 108.

[2] If the object has been preserved in alcohol, it must remain some time in water before imbedding.

[3] This method is recommended by Kaiser. Botan. Centralbl. I., p. 25, 1880.

[4] Morphologisches Jahrbuch, II., p. 445, 1876.

fluid with which the hollow in the solid block[1] is wet, the block itself being only a piece of the same mixture after it has been hardened in alcohol.

Calberla soaks the eggs a few minutes (5-20) in the fresh white of the egg before imbedding; Hertwig appears to omit this part of the process.

After the egg has been fixed to the block, it is placed in a paper box, and covered with the fresh mass (1-2^{cm} deep). The box is then placed in a vessel that contains alcohol (75-80 %) enough to bathe its lower half; the vessel, covered with a funnel, is heated over a water-bath for 30-40 minutes, care being taken not to *boil* the alcohol. The imbedding substance, thus hardened, is next placed in cold alcohol (90 %) which should be changed once or twice during the first twenty-four hours. After remaining in alcohol for about forty-eight hours, the imbedded egg is ready for cutting.

Ruge's Method of making Egg-Emulsion. — According to Scott,[2] Dr. Ruge prepares egg-emulsion in the following manner: —

The yolks and whites of as many eggs as are needed are shaken violently together, with the addition of four drops of glycerine to each egg. The mass is then filtered through fine flannel, which removes all the membranes and air bubbles, and gives a beautifully homogeneous substance. The hardening is done as in Calberla's method, and the alcohol entirely removes the yellow color of the mass, leaving it as white as if the whites alone had been used. This substance becomes perfectly transparent in oil of clove or balsam, and cuts with extreme readiness, allowing very thin sections to be made. The objection to Calberla's method is that the carbonate of soda is apt to crystallize out in long needles, and so injure the preparations.

CELLOIDIN.

Schiefferdecker's Method.[3] — Celloidin is a form of nitrocellulose (pyroxylin) prepared in plates, and may be obtained

---

[1] See Hertwig's method of imbedding the frog's eggs, infra.
[2] Science Record, II., pp. 41, 42.
[3] Arch. f. Anat. u. Phys., I. Abth., p. 199, 1882.

from Wittich & Benkendorf, Chausee Strasse 19, Berlin, N.; from Bachrach & Brothers, Baltimore; or from the Educational Supply Company, 6 Hamilton Place, Boston.

Schiefferdecker uses two solutions, one of a syrupy consistency, the other somewhat thinner. The celloidin plate is cut into small pieces, and dissolved in absolute alcohol and ether (in equal parts). Objects are transferred from absolute alcohol,[1] first to the thinner solution, then to the thicker. After remaining a few hours (or days, according to the character of the object) in the latter, they are imbedded in paper boxes. As soon as a hardened film forms on the solution in the box, the whole is placed in 82 % alcohol for 24–48 hours, and thus rendered sufficiently hard for cutting.

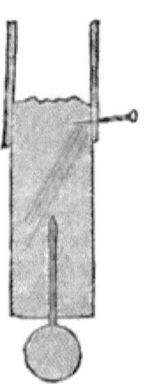

FIG. 27.

Blochmann[2] recommends imbedding on a cork rather than in a paper box, as less celloidin is required, and as the cork is held more firmly in the holder. One end of the cork is made rough, and surrounded by a strip of paper, which is made fast by a pin, as shown in the figure. The roughened surface of the cork is wet with absolute alcohol, and then the object imbedded in the usual manner. In order that this small box may sink in alcohol, in which it is placed for hardening the celloidin, it may be weighted with a small lead ball fastened to the cork by a needle (see figure).

In cutting, the knife is kept wet with alcohol (70 %). The sections may be placed in water or in alcohol, and afterwards stained with carmine or hæmatoxylin, in which the celloidin is only a little, or not at all, stained. Aniline dyes color the celloidin, and therefore should not be used.

The sections can be mounted in glycerine or in balsam; but in the latter case they must be dehydrated with 95 % alcohol, as absolute alcohol dissolves the celloidin. They should be

---

[1] If the objects are penetrated with difficulty, they may be transferred from absolute alcohol to ether, then to the celloidin solutions.

[2] F. Blochmann. "Ueber Einbettungsmethoden." Zeitschr. f. wiss. Mikr., I., H. 2, p. 218, 1884.

clarified in bergamot oil or origanum oil (clove oil dissolves the celloidin).

Objects imbedded in celloidin can be preserved ready for cutting for a long time in 70-80 % alcohol.

Thoma on Methods of Imbedding in Egg-Emulsion and Celloidin. — The method of treating tissues with gum arabic, first brought into use by Rindfleisch and Ranvier, is now very generally known and practised. The same may be said of the method of cutting sections between two pieces of elder pith or hardened liver, etc. These, in certain conditions, are very useful and simple; but other methods of imbedding, of more recent date, give sections of the utmost perfection and unsurpassed delicacy.

The method of imbedding in emulsions containing fat and albumen originated with Bunge, and was subsequently modified by Calberla and Ruge. The following is very nearly the formula of the latter: The albumen and yolk of several hens' eggs is placed in a porcelain mortar, and well stirred until it forms a thin yellow fluid, a result generally obtained in a few minutes. This fluid is subsequently passed through thin linen, in order to remove the remaining membranaceous fragments. The specimen, previously hardened in alcohol, is then fixed by pins in a paper box, and covered with the fluid.

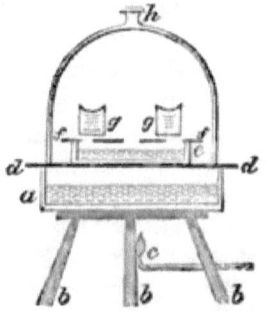

FIG. 28. — Thoma's imbedding apparatus.

The preparation cannot, however, be immersed directly in alcohol for the purpose of hardening. It must be first hardened by alcohol steam, taking care never to raise the temperature of the steam above 30° C. For this purpose, Professor Thoma uses a simple apparatus represented in Fig. 28.

A shallow water-bath, a, stands on an iron tripod, $b\,b\,b$, and is heated by a small flame, c. The water-bath is covered by a thin plate, $d\,d$. Upon this plate is a small glass vessel, e, filled with common alcohol, and covered with a perforated disk of tin, $f\!f$. On this disk are placed the paper boxes, $g\,g$, containing the specimens and the imbedding fluid. The latter and the alcohol vessel are again separated from the external air by a glass cover, *h*. This apparatus, slightly heated, will harden the imbedding masses within a few days, after which time they are removed, and subsequently fully hardened in a bottle containing ordinary alcohol. The latter process determines the degree of consistence of the imbedding mass. It can be made extremely hard by repeated use of strong alcohol. After a few trials it will be easy to find the convenient degree of consistence for each specimen.

If the temperature of the alcohol steam is more elevated, it will be found that the imbedding mass, instead of shrinking, will appear to increase in volume, innumerable air bubbles developing in the emulsion. This can be easily avoided by using lower temperatures. Another danger, however, exists in the holes which the pins make in the walls of the paper boxes. The emulsion before hardening is so very liquid that it will pass through the smallest opening; this renders it necessary not to withdraw any of the pins from the sides of the paper box, and to use boxes without any openings. It will be found that this mass adapts itself perfectly to all surfaces of the specimens without penetrating into their interior structure, and that it can be cut admirably at all thicknesses down to 0.003^{mm}. Another very agreeable quality results from the fact that the newly prepared emulsion will adapt itself readily to hardened pieces. This enables us to spread out fine membranes on pieces of the hardened imbedding mass, and subsequently to imbed both in the way just described.

After this praise of the egg-emulsion, it will be just to mention a property which is occasionally disagreeable. It cannot be easily detached from the sections, and we have no means of dissolving it in media which do not injure the objects. The mass also colors in all the staining fluids generally used, and therefore becomes very visible in the preparations. The latter inconvenience should in all cases be avoided by coloring the specimen *in toto* before imbedding. For this purpose the fluids of Grenacher,[1] and especially alum-carmine, may be recom-

---

[1] Arch. f. mikr. Anat., XVI. (1879), p. 465.

mended. The imbedding mass remains nearly absolutely colorless if the specimen, after staining and before imbedding, is hardened again in alcohol.

Very elegant results may also be obtained by an imbedding mass originally invented by Duval, and recently much improved by Merkel and Schiefferdecker.[1] This is collodion, or, preferably, a solution of so-called celloidin. If this substance cannot in general be cut to such extreme delicacy as the albuminous mass just described, it has a great advantage in being extremely pellucid.

According to the formula of Schiefferdecker, the imbedding fluid consists of a concentrated solution of celloidin in a mixture of equal parts of absolute alcohol and ether. The specimen is soaked successively in absolute alcohol and ether, and in the imbedding fluid. This requires at least several days. After this time, the imbedding proper may be undertaken, and for this we have the choice of two methods.

The even surface of a cork is covered with a thick solution of celloidin, so as to form by evaporation a strong collodion membrane on the cork. Upon this is put the specimen, covered layer by layer with fresh quantities of the solution of celloidin, each being allowed to dry only partially. When the object is thoroughly covered, we immerse it in alcohol of 0.842 sp. gr. In twenty-four hours the whole is ready for cutting.

The other method makes use of little paper boxes for the imbedding. The specimen, soaked in celloidin solution, is fixed in the box by pins, and the box filled with celloidin. The preparation is then placed on a flat piece of glass, and covered with a glass cover which does not exactly fit the glass plate. In a few days the ether will have evaporated gently and slowly from the imbedding mass, and the latter will shrink a little. If necessary, further celloidin solution can be poured into the paper box to fill it again. It is only necessary to moisten the surface of the first mass with a drop of ether in order to allow of a perfect junction between the old and the new layers. The preparation is again exposed to slow evaporation below the glass

---

[1] Arch. f. Anat. u. Physiol. (Anat. Abtheil.), 1882.

cover and a few days later the imbedding mass will be consolidated to an opaline body, whose consistency can well be compared to that of the albumen of a boiled egg. The walls of the paper box can now be removed, and the imbedding mass placed in very dilute alcohol, which will, in a very few days, produce a proper degree of consistency to admit of cutting.

This method differs, in some degree from that which Schiefferdecker gives for imbedding in paper boxes. As other observers have remarked, his method frequently gives rise to a great number of air bubbles in the imbedding mass. Consequent upon the altered manipulations of Professor Thoma, we have to adapt the imbedded specimen to a cork for the purpose of cutting. This may be done in the following way: The even surface of the cork is covered by a thick layer of celloidin solution. This is allowed to dry up perfectly, so as to produce a hard membrane of celloidin. This is again covered with further celloidin solution. In the mean time the lower surface of the imbedding mass is cut even and washed with absolute alcohol, and subsequently moistened with a drop of ether. This moist surface is adapted to the stratum of liquid celloidin on the cork, and exposed for a few minutes to the open air. After this, the whole is placed in dilute alcohol, which in a few hours will unite the imbedding mass solidly with the cork.

In a great number of cases it may be regarded as a great advantage of the celloidin that it penetrates the tissues thoroughly, and yet remains pellucid, so as to be more or less invisible in the specimen. This quality can be made use of in another direction for the purpose of soaking specimens which are too brittle to be cut after hardening alone. We may make use of celloidin in a similar way to the gum arabic mentioned above. The minute normal and pathological anatomy of the lung, in particular, will derive great advantage from such a proceeding. Indeed, we are not able to get a perfect idea of the changes produced by pneumonia, if we do not, by this method, or by the following (with paraffine), prevent the loss of a great part of the exuded substances which in this disease lie loose in the alveolar cavities. The study, also, of micro-organisms in the lung will derive great benefit from the celloidin method, and it will be very welcome to many to know that the tissues imbedded in celloidin may be stained with the different fluids, — ammonia-carmine, alum-carmine, borax-carmine, hæmatoxylin, analine colors, and various others. The reaction of acids and alkalies — particularly acetic acid and solution of potash — is, moreover, not interfered with. And further, we are able to color the object before imbedding with all staining fluids which are not soluble, or only little soluble, in alcohol and ether.

After staining and cutting, the sections may be mounted in glycerine and various other fluids. Mounting in Canada balsam requires, however, some precautions, on account of the chemical character of the celloidin. Absolute alcohol and oil of clove should be avoided, and replaced by alcohol of 96 %, and by oleum origani.

Minot's Method of Imbedding in Celloidin. — The object, after having been thoroughly dehydrated in alcohol, is placed for twenty-four hours in a mixture of equal parts strong alcohol and pure ether. If this mixture is kept long, a little ether must be added from time to time. Transfer to a thin solution of celloidin, and allow it to remain for from one to three days, according to the size of the object.

Imbed in a thicker solution of celloidin. The imbedding box is prepared as recommended by Blochmann. Bubbles will rise from the cork, and interfere with the imbedding. Two precautions will essentially diminish this danger : 1. Pour in so much celloidin that it covers the object half an inch deep, giving an opportunity for the bubbles to rise above the tissue; 2. Before imbedding, cover the end of the cork with a thin layer of celloidin (as recommended by Thoma), which is allowed to dry on completely. After the object is covered, the cork is mounted on a lead sinker, and allowed to stand until a film has formed on the upper surface. It is then immersed in alcohol of 82–85 % (stronger alcohol attacks the celloidin) for one to three days. I have found it best to allow plenty of time for the hardening after imbedding.

The sections have to be cut under alcohol. We use Jung's microtome with his largest knife placed so as to cut with as much of the blade as possible. If the edge is good, then the longer the draw, the thinner the section which can be made. While cutting, the knife-blade should have as much alcohol upon it as possible; to secure this, we use the dripping apparatus described below. The sections should be removed from the knife with a fine brush to avoid all risk to the edge. For celloidin imbedding are needed, —

1. Mixture of ether and alcohol, equal parts.
2. A thin solution of celloidin in (1). This solution should be syrupy, but still flow easily.
3. A thick solution of celloidin in (1) of about the consistency of thick molasses. The usual mistake is to have the solution too thick. Quantitative directions cannot be given, because the celloidin varies in weight, according as it is more or less dried.
4. Alcohol, 82–85%.

For mounting sections with celloidin left on them, I have found none of the methods hitherto recommended satisfactory. The essential oils I have tried either dissolve the celloidin, like oil of clove, or cause it to shrink and distort the section, like oil of bergamot. After trying various reagents, I have settled upon chloroform as the most convenient medium of transfer from alcohol to balsam. In using it care must be taken to place the section for half a minute in perfectly fresh alcohol, which is really 95–96%; — if this is done, chloroform will clear it up almost immediately. When the section is in chloroform on the slide, the mounting must be expeditious, and the balsam added *while the chloroform is still covering the section.* If many sections are to be mounted, it is convenient to have a dish full of chloroform, and large enough to permit plunging in the slide and placing the section on it, all under chloroform. The transfer, particularly of a large section, from the spatula to the slide with chloroform, is often *very* difficult. To mount a single section, put it in alcohol on the slide, wash with a few drops of fresh strong alcohol; let most of the alcohol drain off, but while the section is still covered with it, add chloroform, drain off the mixture, and pour over the still moist section a fresh dose of chloroform; if the washings have been really thorough, the section will clear at once.

Dripping Apparatus for Cutting under Alcohol. — We use the form constructed by D. W. W. Gannett, as shown in the sketch. A litre bottle is convenient in size; the height of the stand should be such as to bring the end of the dripping-

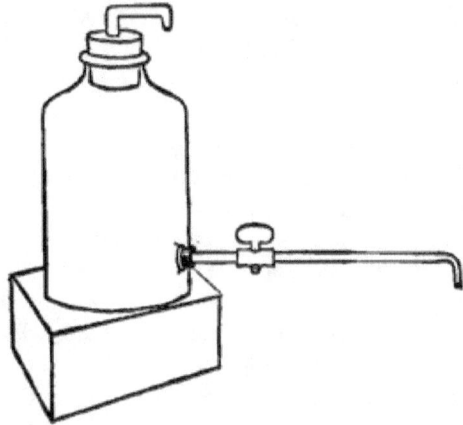

FIG. 29 — Dripping apparatus.

tube about one inch above the blade of the microtome knife, on which the alcohol is allowed to fall continuously. To regulate the flow, an $\frac{1}{8}$ in. globe valve is found to be the most convenient.

CHAPTER V.

FIXATIVES FOR SERIAL SECTIONS.

ALTHOUGH the first attempts at fixing sections, by Mayer and Gaule,[1] were partially successful, the problem was only actually solved by the discovery of *liquid fixatives*. It is to Dr. Giesbrecht, of the Naples Zoological Station, that we are indebted for the first fixative of this kind. This method, still widely used, has recently been improved by a new method of dissolving the shellac, discovered by Mayer. The collodion method of Schällibaum, is perhaps now more generally employed, as it enables one to stain *after* sectioning, and serves equally well with the shellac method for sections of objects stained in toto. Mayer's albumen fixative offers some advantages. The sections can be treated with turpentine, alcohol, water, staining fluids, glycerine, etc., without danger of loosening. In staining nuclei, the film of albumen remains uncolored, which is an advantage over the collodion method.

Giesbrecht's Shellac Fixative.[2]—The shellac is prepared and used in the following manner. One part of bleached shellac[3] dissolved in ten parts absolute alcohol and filtered. The object

---

[1] Francotte. "Description des différentes méthodes employées pour ranger les coupes en séries." Bull. de la Soc. belge de microscopie, X; No. 2, p. 43.

[2] Giesbrecht. "Methode zur Anfertigung von Serien-Praparaten," in Mittheilungen a. d. Zoolog. Station zu Neapel, 1881, p. 184.

[3] The preparation of bleached shellac used by artists as a "fixative" for charcoal pictures, as Dr. Mark first suggested, may serve equally well.

glass is first warmed to about 50° C.,[1] and then a thin film of the shellac is laid on by a glass rod drawn once or twice over its surface. Before using, the slide is again warmed, and the shellac surface washed with oil of cloves for the purpose of softening it. The wash is made with a small brush drawn back and forth until the entire surface has been moderately but evenly wet with the oil. Sections are now cut and arranged for the first cover; this done, the slide is warmed over a spirit lamp, so that the paraffine adhering to the sections melts and flows together, forming an even layer, which cools almost instantly, and thus secures the position of the sections, while those of the second cover are prepared. The sections for the last cover having been completed, the slide is warmed for ten minutes on a water-bath, in order that the sections may sink into the shellac and become fixed, and the clove oil evaporate. After allowing the slide to cool, the process is concluded by washing away the paraffine with turpentine, and inclosing in balsam dissolved in chloroform.[2]

Method of Fixing Collodion Sections. (Minot.) — Sections of objects imbedded in celloidin are "fixed" in serial order, preparatory to mounting, in the following manner :—

The sections are arranged on the slide in 95% alcohol; then the alcohol is poured off, and a drop of alcoholic shellac placed on each section (just enough to cover the section completely). The slide is next placed in the oven of a water-bath at 40°, for a few minutes (5-10), until dry. The sections are then ready for clarifying in clove oil, and mounting in balsam.

Mayer's Carbolic Acid Shellac. — The use of clove oil, or creosote in the shellac method of Giesbrecht, leaves fine granulations on the slide. In order to avoid this, Dr. Mayer proposes to use a solution of shellac in carbolic acid. The following is Mayer's method of preparation, to which I have added only a few details.

---

[1] The same temperature is used throughout the operation.
[2] Occasionally one experiences great difficulty in dissolving the bleached lac of the stores. This is due to the fact that now the bleaching is accomplished by means of chemicals. One can, however, be assured of success (we are informed) by bleaching the lac for themselves, by exposure to the sun. (Kingsley.)

1. Dissolve 50 g. bleached shellac in 250 g. absolute alcohol.

It is well to powder the shellac and leave it 24 hours or more in the alcohol, shaking it from time to time. Only a small portion of the shellac dissolves. Probably a much larger amount would dissolve by heating, but I have not tested this.

2. Filter the solution, and evaporate the alcohol on a water-bath.

The filtering paper should be wet with absolute alcohol when the solution is poured into it. The evaporation should continue until the yellowish residue, when cold, is quite stiff.

3. Dissolve shellac residue in pure carbolic acid, on a water-bath, using from 3 to 5 volumes of acid to one of shellac. The solution, after cooling, should be nearly as thick as glycerine.

It is important to use a concentrated solution of carbolic acid, and this may be obtained by exposing the crystals to the air until they dissolve, or by adding about 5% water, and warming.

Following these directions, I obtained about 30 grams of the fixative.

I keep the fixative in a bottle covered with a cap, like that of a spirit-lamp. A small brush is thrust through a flat piece of cork, which rests on the mouth of the bottle, and allowed to dip barely below the surface of the liquid. If the brush is thrust deep into the fixative, the hairs curl up, and the brush is otherwise liable to be affected by the acid.

The shellac can be directly dissolved in carbolic acid, but then the liquid must remain standing a long time in order to become perfectly clear, for it cannot be filtered. For this reason, I prefer, with Mayer, to make the solution first in alcohol.

Mayer now prepares as follows: —

The shellac is first pulverized, then heated with colorless crystals of carbolic acid until it dissolves. It is next to be filtered, the funnel being kept hot during the process, by heating over a flame.

This fixative, which is much stronger than any alcoholic solution I have been able to obtain, is applied to the *cold* slide with a small brush, at the time of using. The brush should be wiped on the mouth of the bottle in order to remove most of the fixative, and then the middle portion of the slide painted with a single thin coat. After the sections have been placed, the slide is placed in the oven of a water-bath, kept at about 55° C., and left from 15 to 20 minutes. The carbolic acid evaporates, leaving a perfectly transparent stratum of shellac.

The sections are then freed from paraffine in the usual way, and mounted in balsam.

GAULE'S METHOD OF FIXING SECTIONS.[1]

1. Sections cut dry and placed on the slide in the order and position in which they are to be mounted.

2. They are then smoothed out by the aid of a fine brush wet in 50–60 % alcohol, until all wrinkles are removed and every part is in close contact with the slide.

3. Slide allowed to stand several hours (or over night) until the alcohol has completely evaporated, and the sections are left adhering quite firmly to the glass. The process may be hastened by gently warming to 45–50° C.

4. The paraffine may be removed by any of the solvents in common use, but Dr. Gaule recommends xylol. A few drops are allowed to flow over the sections, and after a few moments the paraffine is fully dissolved.

5. The balsam (a mixture of balsam and xylol in equal parts) is placed on the cover-glass, and this allowed to sink slowly, from one side, over the sections.

Dr. Gaule finds it convenient, especially with serial sections, to use large cover-glasses, often nearly as large as the slide itself. Thus a single slide may often contain a large number of sections closely arranged under one cover.

For large sections, this method offers one important advantage over that of Dr. Giesbrecht; for by the former all wrinkles may be removed, while by the latter the sections must lie as they fall. In the case of smaller sections, not liable to get wrinkled during the placing, I prefer the shellac method.

FRENZEL'S GUTTA-PERCHA FIXATIVE.[2]

The only advantage claimed for this fixative is that it allows the sections to be colored after they are fixed on the slide. The difficulty in dissolving the paraffine from the sections is the chief objection to Frenzel's method, which is as follows: —

A solution of gutta percha in chloroform and benzine is used.

---

[1] Arch. f. Anat. u. Phys., 1881, Phys. Abthl., p. 156.
[2] Zoologischer Anzeiger, No. 130, p. 51, 1883.

The solution must be filtered until it is clear and almost colorless, and must not be made so thin that it will flow quickly on the slide. The fluid is put on with a brush, and allowed to dry before the sections are placed. If the object is imbedded in paraffine or a paraffine mixture (*e. g.*, four parts paraffine and one part vaseline), the sections are smoothed out by the aid of a brush wet in alcohol, and then exposed to a temperature of 35° to 50° C. for five to ten minutes, in order that the gutta percha may become sticky; they are then left for five to ten minutes in the air, and finally placed in a vessel containing absolute alcohol heated to 40–50° C., in order to remove the paraffine. Five to fifteen minutes are required to dissolve the paraffine, and a considerable quantity of alcohol must be used, as it is a poor solvent of paraffine. After the alcohol becomes saturated, it may be filtered cold, and used again. After the paraffine has been completely removed, the sections are transferred to 70 % alcohol, then to lower grades, and finally to water. The sections are then ready for staining. After staining and washing, they are soaked in absolute alcohol, and thus prepared for mounting in clove oil and balsam.

If the object is imbedded in celloidin, as is now frequently done, the sections may be smoothed out with benzine or chloroform, which softens the gutta percha, and thus fixes them in position. After the sections have dried on, they may be stained, washed, and transferred to absolute alcohol as before. The application of clove oil, before the balsam, dissolves the celloidin.

THRELFALL'S CAOUTCHOUC FIXATIVE.[1]

A thin solution of caoutchouc in benzine or chloroform is prepared and poured over the slide so as to form a film, in the same way that collodion is poured on a photographic plate. When the film is dry, the sections are arranged on it, and the temperature of the slide raised to the melting-point of paraffine; the sections then fall on to the India rubber film, which has become sufficiently sticky to adhere to them perfectly. *When the slide is cold*, it is treated with náphtha, or any light paraffine oil,

---

[1] Zool. Anzeiger, No. 140, p. 300, 1883.

the solvent action being more rapid the lower the boiling-point of the oil used.

Absolute alcohol is readily miscible with the naphtha or light paraffine, so that the solvent is readily removed. The slide can now be placed in successive alcohols, stained, and returned to absolute alcohol. It is now to be cleared with creosote or oil of clove, and mounted in the ordinary way.

SCHÄLLIBAUM'S COLLODION FIXATIVE.[1]

The solution, which is prepared by dissolving *one part collodion in three or four parts clove oil*, is applied to the slide by means of a fine brush, at the time of using. The sections having been arranged, the slide is warmed for a few minutes (5–10) in the oven of a water-bath, in order to evaporate the clove oil. The sections may next be freed from the imbedding mass, and colored according to desire. If the film of collodion be too thick, cloudiness is likely to arise between the sections. The cloudiness can be removed by the use of a brush wet with clove oil, after the sections have been dehydrated by absolute alcohol.

Gage,[2] who had begun to experiment with collodion before Schällibaum's method was published, recommends that the collodion and clove oil be applied separately: —

"A solution of collodion is prepared by adding to 2 grams of gun cotton (that used by photographers is good) 54^{cc} of sulphuric ether and 18^{cc} of 95 % alcohol. After the gun cotton is entirely dissolved, the solution should be filtered through filter paper or absorbent cotton. The slides are coated by pouring the collodion on one end, allowing it to flow quickly over the slide and off the other end into the bottle. The prepared slides should be kept free from dust. As the collodion will not deteriorate after drying on the slide, any number of slides may be prepared at the same time. Before using a slide, it should be dusted with a camel's-hair brush; and with another brush the collodionized surface of the slide should be thinly painted with clove oil.

---

[1] Arch. f. mik. Anat., XXII., p. 689. 1883.
[2] H. Gage and T. Smith. The Medical Student, p. 14, Nov., 1883.

"... The sections are arranged as in the shellac method. The slide is warmed over an alcohol lamp and then heated in a warm chamber so as to drive off the clove oil. After cooling, it may be placed in a wide-mouthed vial of turpentine, chloroform, xylol, or refined naphtha, to remove the paraffine. Naphtha is very cheap, and is the best agent we have yet tried for dissolving **the** imbedding mass. The sections are usually freed **from the** imbedding mass within half an hour, though the slide **may remain in** any of the solvents mentioned for two or three **days,** or perhaps indefinitely, without loosening the sections. **When the slide** is removed from the naphtha, the sections are **washed with 95 % alcohol by means of** a medicine-dropper, or **by immersing the slide in alcohol.** If the sections are to be **stained in Kleinenberg's** hæmatoxylin, or in any other stain **containing 50 % or more acohol,** the slide is transferred directly **from the alcohol used** for rinsing to the staining agent, otherwise **it should be first** transferred to 50 % alcohol, and from that **to the staining agent.** Whenever the sections are sufficiently **stained, they may be mounted in any desired** mounting medium. In **case Canada balsam is to be used, the slide** must be immersed **in alcohol to** wash away the stain, and finally in 95 % alcohol to completely dehydrate the sections. They are cleared with a **mixture of** carbolic acid, 1 part; turpentine, 4 parts. The bal**sam to be used** is prepared by mixing 25 grams of pure Canada **balsam with 2^{cc} of** chloroform and 2^{cc} of clove oil. The latter **very soon removes any** cloudiness that may **have** appeared in **the collodion film."**

MAYER'S ALBUMEN FIXATIVE.[1]

The new fixative proposed by Dr. Mayer is prepared by mixing *the filtered white of eggs with an equal volume of glycerine*.[2] A little carbolic acid may be added as an antiseptic. A very thin and even layer of the fixative is painted upon the object-slide, which is then ready for receiving the sections. After the

---

[1] Mitth. Zoolog. Station zu Neapal, IV., p. 521, 1883.

[2] Glycerine serves to keep the fixative moist while the sections are being placed.

sections are placed, the slide is warmed[1] a few minutes in the oven of a water-bath, just long enough to melt the paraffine. The paraffine is next dissolved away from the sections by turpentine, the turpentine removed by alcohol, and the sections then colored *in situ* on the slide.

As a coloring fluid, a strong alcoholic carmine is recommended. It is prepared as follows: —

Dissolve four grams carmine in 100^{ccm} alcohol (80 %) by boiling about thirty minutes, adding thirty drops of strong hydrochloric acid during the boiling. The solution should be filtered while hot, and the acid carefully neutralized by adding ammonia until carmine begins to be precipitated.

This fluid, which may require to be filtered a second time after cooling, stains uncommonly quick, deep, and diffuse. To obtain a differential staining, it is well to stain deeply, and then partially decolor by washing with acidulated alcohol.

The decoloration should be checked at the moment when the film of the fixative becomes nearly or quite colorless.

This fixative may be employed to great advantage in the process of staining with aniline dyes (*vide* Flemming's Methods of Treating Nuclei). After the sections have been fixed on the slide and colored with some aniline dye, the slide may be washed quickly in 70 % alcohol, then placed in 95–96 % alcohol for a few moments *until the fixative has lost nearly all its color*. The decoloration of the fixative can be followed so readily that it seems to mark the precise moment when the process of decoloring should cease.

---

[1] Care should be taken not to cause coagulation of the film by overheating.

CHAPTER VI.

MOUNTING MEDIA.

Although the preservative and optical qualities of glycerine make it indispensable in many ways as a mounting medium, still the inconveniences attending its use for permanent histological preparations and its evident non-adaptability to the needs of the microtomist have tended to give balsam and damar the preference as mounting media.

There has been some difference of opinion on the question whether Canada balsam or damar varnish is the better mounting medium. It is generally known that uncolored structure is best seen in damar, and that balsam clarifies more strongly than damar. It is not, therefore, advisable to limit one's self to either medium exclusively, but rather to make a selective use of both.

Balsam is usually dissolved in turpentine, xylol, benzole, or chloroform. Benzole, when used as a solvent of balsam, should be chemically pure and *free from* water, otherwise very fine, drop-like formations arise, which give the preparation a cloudy appearance to the naked eye. Such results are most likely to appear in preparations mounted at the seashore. Such preparations can be restored by placing the slides in a bottle with benzole enough to cover them. The bottle should be laid on its side, and the slide inverted so that the preparation may fall freely, and thus escape crushing by the cover-glass. In the course of an hour or two the preparations are set free, and may then be transferred to clove oil and remounted in balsam.

Canada balsam is prepared by heating for several hours over a sand-bath, until it is quite hard when cold. It is then dissolved in an equal volume of one of the above solvents. The balsam should be so hard before dissolving it that it will break like glass when cold.

Damar may be prepared by dissolving the gum in pure turpentine or in a mixture of turpentine and benzole in equal parts. This method is preferable to that given by Schäffer, according to which —

One part of gum damar is dissolved in two parts of oil of turpentine, and one part of gum mastic in two parts of chloroform. The two solutions are mixed and filtered. Schäffer states that the opalescence which often arises after a time disappears permanently on boiling the fluid.

GLYCERINE JELLY.

The two following solutions, similar in composition to Bachmann's, are employed by Fol.

1st. Gelatine . . . . 30 parts.
    Water . . . . 70 "
    Glycerine . . . . 100 "
    Camphor (alcoholic solution) . . . . . 5 "

2d. Gelatine . . . . 20 parts.
    Water . . . . 150 "
    Glycerine . . . . 100 "
    Camphor . . . . 15 "

The solutions are prepared by first softening the gelatine a few hours in the water, then melting on a water-bath. The glycerine is next added, and finally the camphor solution by drops while stirring briskly.

For use, the jelly is to be warmed *en masse*, or a piece may be cut out and melted on the slide. Over-heating will produce bubbles, which must be carefully avoided. The cover-glass must be encircled by some cement, otherwise evaporation will give rise to air bubbles.

CHAPTER VII.

THE USES OF COLLODION.

In modern histological technique, collodion has come to serve a variety of important purposes. Duval[1] was the first to call attention to its advantages as an imbedding mass. He found that it penetrated preparations easily and thoroughly; that it could be quickly brought to the proper degree of hardness in alcohol of 36° (80 %); that objects thus imbedded could be preserved in this alcohol for an indefinite length of time; that the imbedding mass preserved its transparency, so that the preparation could be easily examined; that the sections did not require to be freed from the mass, since they could be colored and mounted in glycerine, and the mass remain unaffected by the process.

As soon as Duval's discovery became known, Merkel and Schiefferdecker[2] began to experiment with collodion, and greatly improved and extended its use. It was found desirable, first of all, to be able to vary the concentration of the collodion, an end very conveniently reached by Merkel through the use of a solid preparation called *celloidin*, which he dissolved in absolute alcohol and ether in equal parts.

Duval mounted sections of objects imbedded in collodion in glycerine, and was at first unsuccessful in his experiments with balsam. Schiefferdecker has shown that by dehydrating the sec-

---

[1] Journ. de l'anat. et de la physiol., XV., p. 185, 1879.
[2] Arch. f. Anat. u. Physiol. Anat. Abth., p. 199, 1882.

tions with 95 % alcohol, and clarifying in oil of origanum or oil of bergamot, the sections could be mounted in balsam.[1]

Some improvements of minor importance in the process of imbedding have been made by Thoma, Blochmann, and others.

The importance of collodion in microtomy was much increased by the important discovery that in combination with clove oil it could be used as a fixative for serial sections, and that the latter could be colored after they had been arranged and fixed on the slide. This invaluable method, discovered by Schällibaum,[2] presents all the advantages of the shellac method of Giesbrecht, and offers, at the same time, the best means of meeting the difficulties of staining objects *in toto*. The only other fixative, thus far known, which claims to accomplish similar results, is that introduced by Mayer (*v.* Fixatives, Chap. VII.).

Prof. Gage,[3] who began to experiment with collodion as a fixative prior to the publication of Schällibaum's method, has given some valuable directions respecting its preparation and application. Gage applies the collodion and clove oil separately, first coating a number of slides with collodion, which is poured on to one end of the slide, and allowed to flow quickly over it and off into the bottle, and then at the time of using adding a wash of clove oil. In order to remove any cloudiness that may arise in the collodion film, a little clove oil is added to the balsam.

The use of collodion to prevent the crumbling of brittle sections was discovered independently by Duval[4] and Mason.[5] The same method was employed in Semper's Laboratory by Timm,[6] Will,[7] Sharp, and others; and Mark has found it indispensable in sectioning the ova of Lepidosteus. Mason applied the collodion by means of a fine brush, taking up a small drop

---

[1] Celloidin is a good imbedding mass for injected objects that will not bear heating, and especially for such large objects as the vertebrate brain.
[2] Arch. f. mik. Anat., XXII., p. 689, 1883.
[3] The Medical Student, p. 14, November, 1883.
[4] Société de Biologie, 1880.
[5] American Naturalist, 1880, p. 825.
[6] Sempers Arbeiten, VI., p. 110, 1883.
[7] Sempers Arbeiten, VI., p. 7, 1883.

and placing it "in the centre of the object, so as to allow it to flow out on all sides, to prevent the formation of air bubbles. After being allowed to harden a minute, the section may be cut and placed on the slide *with the film of collodion underneath.*" Mark and others who have used collodion for the same purpose simply paint the cut surface of the object with a thin film a few seconds before making each section.

Celloidin has also been recommended by Schiefferdecker for forming injection masses. It penetrates easily, and its viscosity protects the vessels from injury in the process of making corrosion preparations (*v.* Methods of Injection).

Part II.

SPECIAL METHODS.

CHAPTER VIII.

EMBRYOLOGICAL METHODS.

DRAWING APPARATUS.

In Part I. of his "Anatomie menschlicher Embryonen," pp. 8, 9, Professor His has described a drawing apparatus similar to the one here represented.

For anatomical and embryological work, an apparatus of this kind is simply indispensable. As every working naturalist knows, an apparatus that admits the use of the *camera lucida* with a low magnifying power, varying from five to forty diameters, offers many advantages that are not to be obtained from any system of microscopical objectives. In the absence of such an instrument, one is compelled to draw by measurement and "by the eye," a process which at best is slow and tedious, and liable to many inaccuracies. The foundation of every thorough embryological work consists, as Professor His remarks, *of exact drawings of the entire embryos as well as the sections obtained from them.* Any one acquainted with the embryometrical investigations of Professor His on the chick, will hardly require to be told that such surface views as he employed for orientation in microtomic sections could not be obtained without the aid of photography, or the camera lucida, or by both. The instrument here described offers the same facilities for obtaining the exact topographical relations of a complicated object with low magnifying powers that the microscope affords with higher powers. Further, only a single plano-convex lens 2.5^{cm} in diameter, is required for an enlargement varying at pleasure from five to fifty diameters. Professor His employs as an objective a stereoscope-head (of Dallmeyer), or a small Steinheil aplanat (No. 1).

The instrument consists of a heavy circular iron base,[1] from the centre of which rises a brass rod, marked to centimeters, half centimeters, and millimeters. On this rod are seen the mirror (M), the object-table (T), the objective (O), and the camera lucida (P), all supported by horizontal bars that move on sliding ferrules. The mirror is placed as near the base as convenience will allow, and its supporting bar is 7.5^{cm} long. The bars bearing the other pieces are all of corresponding length, and the sliding ferrules can be fixed at any point by the aid of setscrews. The ferrules of the mirror and the object-table are made of such length that when in contact with each other and resting on the highest part of the base, they are in the position required for the lowest magnifying power. In this position the object-table has an elevation of 11^{cm}, the objective 18.5^{cm}, and the camera 22^{cm} above the lower face of the base, or what may be called the drawing plane.

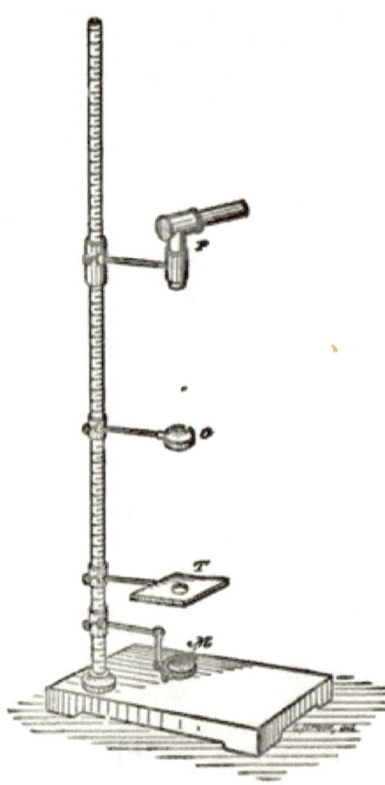

FIG. 30. — Drawing apparatus.

Thus placed, the focal distance is 7.5^{cm}, and the camera is considerably lower than it is possible to have it in ordinary microscopes. The magnifying power is 10 diam., but it may

---

[1] This form has been found more convenient than the rectangular form seen in the figure.

be reduced to 5 diam. by elevating the drawing plane. It is of course possible to obtain a lens that will magnify only 4 or 5 diam. in the normal position of the drawing plane, and endeavors will be made to provide the instrument with such a lens.

The magnifying power may be increased in several ways, but most conveniently by shortening the focal distance and raising the camera until it is properly adjusted. Starting with the parts placed as above given, in order to raise the magnifying power from 10 to 15 diam., the objective must be lowered 7^{mm} and the camera elevated 2^{cm}; and for 20 diam., the objective must be lowered 10^{mm} and the camera elevated 3.5^{cm}. Keeping the object-table in its first or normal position, the magnifying power may be increased to 50 diam. by lowering the objective 2^{cm} and raising the camera 15^{cm}. The camera thus has an elevation of 37^{cm} above the normal drawing plane. The magnifying power may also be increased by lowering the drawing plane, or what is the same thing, by raising the object-table.

The positions which the objective and camera must have for a given magnifying power will probably vary slightly in two instruments made as nearly alike as possible, but their determination is a very simple matter; and having been once accurately ascertained, they may be tabulated and kept for subsequent use. To ascertain these positions for any given magnification, a millimeter scale may be placed on the object-table, and the camera and objective moved until the picture projected on the drawing plane has the desired enlargement. When the scale is replaced by the object, care must be taken to have the surface which is to be outlined in the plane previously occupied by the scale. To this end it may be necessary to move the object-table a very little, in order to give a sharply defined picture, the positions of the camera and objective being left unaltered.

The object-table measures 8×10^{cm}, and has a central perforation 2.5^{cm} in diam. The whole apparatus is completed by a movable shade, designed to cut off the light falling on the lens and on the drawing plane.

It is hardly necessary to remark that opaque objects require direct sunlight or light from a lamp supplied with a bull's eye condenser.

This instrument, including lens, and Oberhäuser's camera lucida, may be obtained from The Educational Supply Co., 7 Hamilton Place, Boston, for thirty dollars. For everything except the camera, the price is fifteen dollars.

MR. IIJIMA'S METHODS OF PREPARING PLANARIANS AND THEIR EGGS.[1]

In the preparation of Planarians for histological study, Mr. Iijima recommends corrosive sublimate as the only good preservative agent. The worms are placed in a shallow plate, *without water*, and a saturated solution of corrosive sublimate, heated almost to boiling, is poured over them. In this way they are killed so quickly that they do not have time to contract. They are left thirty minutes or less in the sublimate; then placed in water for an hour or more. The water should be changed several times, in order to remove all of the sublimate; otherwise it forms needle-like crystals, which impair or ruin the preparation. Three grades of alcohol ("weak, strong, and absolute") are used in hardening, in each of which the object should be left at least forty-eight hours before staining. Borax-carmine (probably the alcoholic solution) is recommended as a staining agent; a dilute solution is used in preference to the full strength, and allowed to act from three to four days.

For preservation as museum specimens, they are killed with strong nitric acid (about 50 %), in which they die fully extended.

Preparation of the Ova.— The egg-capsules of fresh-water Planarians are generally attached to water-plants by means of a white secretion. The ova are very small and few in number, and are scattered among an immense number of yolk-cells. The ova are completely naked, and a little smaller than the yolk-cells, and are not easily isolated. When cleavage begins,

---

[1] "Entwicklungsgeschichte der Süsswasser-Dendrocœlen." Zeitschr. f. wiss. Zoologie, XL., p. 359, 1884.

a large number of yolk-cells surround the ovum, and form with it a mass large enough to be seen with the naked eye. Mr. Iijima adopts the following mode of isolation and preparation: By the aid of two sharp dissecting needles, the egg-capsule is opened on an object-slide in dilute acetic acid (2 %). The contents flow out, and the empty capsule is then removed. The slide is next shaken in order to isolate the ova so far as possible from the yolk-cells. This process detaches many of the yolk-cells, but not all; each ovum will still have yolk-cells adhering to it, and will now appear to the naked eye as a minute white mass. A cover-glass, supported by wax feet or by slips of paper, is now placed over them. After about thirty minutes, the acetic acid is carefully removed by the aid of small pieces of blotting paper placed at one side of the cover, and replaced by alcohol (70 %). The withdrawal of the acetic acid must be as slow as possible, otherwise the ova will be lost. After an hour the alcohol is replaced by a stronger grade (90 %), in which the ova should remain two hours. Finally, the alcohol is replaced by a mixture of glycerine and water in equal parts, and this in turn by pure glycerine. The preparation is now complete, and the cover-glass may be fixed in the usual way by means of lac.

In order to obtain sections of embryos which are too small to be treated individually, the contents of the capsule may be hardened *in toto* in chromic acid (1 %), which renders them less brittle than corrosive sublimate.

The changes which take place in the ovum initiatory to cleavage are very difficult to trace, as they are generally completed before the cocoon is laid. In some cases ova were found in fresh-laid capsules, which showed the germinal vesicle still unchanged; others were found to have two nuclei, supposed to be derivatives from the first cleavage-nucleus. This stage of two nuclei was also found in some cocoons taken directly from the penial sheath, in which the cocoon formation takes place. It is therefore not quite certain when fecundation takes place, whether in the cocoon or before its formation.

THE EGGS OF THE BRACHIOPODA.

The eggs of *Terebratulina septentrionalis* are found, according to Morse,[1] throughout the summer months.

Specimens may be collected at low-tide mark, and by dredging at various depths. Morse found them abundant at Eastport, Me.

For the following on methods of obtaining the eggs I am indebted to Dr. Wilson:—

Sexual Distinction.—The sexes are distinguished by the color of the pallial sinuses as seen through the valves of the shell, the males being white, the females yellowish or cream-colored. The eggs are usually discharged at night, probably in the morning hours, and may be obtained in abundance by keeping males and females in about equal numbers (20-30 each) in a vessel of water over night. The jars were kept in a dark, cool cellar in plates of water, and wrapped in cloth. The cloth soaks up the water, and the evaporation keeps up a uniformly cool temperature. Thus kept, the water requires to be changed only once a day.

The eggs will be found at the bottom of the jar, and may be gathered with a dipping tube.

By tearing open the valves of females, embryos may also be often found in two whitish masses at the side of the body behind the mouth.

In the active, free-swimming stage the water cannot be changed, but the larvæ appear to flourish.

THE OVA OF CLEPSINE.[2]

The eggs of Clepsine are in many respects very favorable objects of study. They are easily obtained in abundance, can be kept alive through all stages of development without difficulty, and are large enough to permit of following the developmental changes with a magnifying power of 20-30 diameters. In order to observe the earliest changes, it is not necessary

---

[1] Mem. Bost. Soc. Nat. Hist., II., p. 30, 1869.
[2] Whitman. Quart. Journ. Micr. Sci., XVIII., 1878, pp. 2, 3, 9-11.

to watch for the deposit of the eggs; for I have found that eggs taken from the ovary at the time they are about to be laid develop in the normal manner, so that material may be had at any moment by simply cutting a leech with ripe eggs in two in the middle. These are rare advantages in the study of the phenomena of maturation and cleavage. In addition to all this, certain extremely interesting phenomena display themselves at the two poles of these eggs, — phenomena which have no parallel, so far as at present known, in the rest of the animal kingdom; and all the principal types of cleavage are exhibited in rapid sequence, beginning with the *regular*, and then shifting at successive stages into the *unequal, partial*, and *superficial* (or plasmodial).

Habitat. — In ponds, and in brooks and rivers where the current is very slow, on the under side of stones, pieces of wood, dead leaves, fallen branches, and other objects lodged in muddy places.

Food. — Most of these animals require only one or two meals a season, and, beyond this, demand only clean water. Some suck the blood of fishes, others live on the blood of tortoises. Many of the smaller species derive their sustenance from fresh-water snails, and one (Japan) lives habitually on Anodonta.

Time of Deposit. — *C. sexoculata* lays in April; the allied species of the Eastern States the first week of May. *C. marginata* lays from the 1st of May to the middle of August, while a closely allied species of Massachusetts deposits early in May. *C. bioculata*, a less favorable species for the study of the development, produces eggs from April to September.

Egg-Sacs. — In most species, the eggs are deposited in thin, transparent sacs (1-8); but in *C. marginata* they are covered with a transparent secretion, which hardens in the course of a few minutes, and thus binds the eggs together and to the object on which they are placed.

Preparation for Sectioning. — The best results were obtained from eggs treated with osmic acid ($\frac{1}{10}$ %) 15–30 minutes, then, after passing through the different grades of alcohol, stained *in toto* with Beale's carmine.

Hoffmann recommends bichromate of potassium.

Surface Views of the Germ-Bands. — The linear arrangement of the cells composing the germ-bands is well seen in eggs hardened in chromic acid ($\frac{1}{3}$ %) 5–10 hours.

For views of the inner surface, it is necessary to free the germ-bands from the yolk. The fresh embryo is placed in a drop of water on an object-slide; a little acetic acid (as much as will adhere to a needle) is added, and all placed under the dissecting microscope. With a pair of sharp needles a rupture is made along the dorsal side. By careful manipulation of the needles, the embryo, in most cases, can be led away from the yolk, and stretched out on the slide.

After partially removing the water by the aid of a bit of filtering paper, a few drops of osmic acid are added, with care not to disturb the object. At the end of an hour, the embryo is washed and stained with Beale's carmine. It is again washed, and, after dehydration, mounted in balsam or glycerine. During all this, the object is not once removed from the slide.

The Vascular System. — The course of the blood-vessels cannot be fully made out in adult specimens. For this purpose it is best to take very young individuals (10–15 days old), in which the pigment is little developed.

The live-box enables one to apply pressure sufficient to render distinct the main channels and branches with a low magnifying power.

SPIRORBIS BOREALIS.

The adult Spirorbis lives in a small, coiled, calcareous tube, which has the form of a round, flat disk. These little white coils are found in great numbers attached to seaweed and other foreign bodies. They are easily kept in a small aquarium, and the young may be reared in multitudes. The cleavage stages and earlier larval stages are passed before the tube of the parent is abandoned, and this renders it possible to make a complete study of the development. The necessity of breaking the shell of the adult to obtain the eggs is a drawback.

Fewkes[1] placed the shells in a watch-glass, and crushed them

---

[1] J. W. Fewkes, Am. Nat., XIX., March, 1885, p. 247.

with a spatula. On removing the fragments of the case, the reddish-brown eggs are found in short strings containing from one to four rows of from ten to fifteen eggs each.

Mr. Agassiz[1] states that the development begins in the body of the parent before the egg-sacs are deposited in the cavity of the tube, where most of the stages are completed. Fewkes did not find the early stages of cleavage, but always found the later stages outside the body of the parent.

MYZOSTOMA.

The reproduction of this parasitic chætopod, two species of which are found on Comatula, is not limited to any particular season of the year, — at least in Naples, according to Mr. Beard. Its food is limited to some portion of that of its host, and consists of infusoria, algæ, and other minute organisms.

In studying the development, Mr. Beard[2] obtained his results mainly by examining the different stages in a living condition.

To obtain naturally fertilized ova, a number of Comatulæ infested with full-grown *M. glabrum* are taken, and the arms cut away not far from the disc. The Comatulæ, with their parasites, are then placed in small, deep glasses filled with seawater, through the upper part of which a gentle current is maintained. The ova, which are laid nearly every day, sink to the bottom, and may be readily found by removing the Comatulæ, and pouring off most of the water. The ova may be kept alive four or five days after removing the Comatulæ.

Artificially fecundated ova may be easily obtained in abundance in the following manner: A number of mature Myzostoma are carefully removed from their hosts, and placed in a watch-glass holding two or three teaspoonfuls of clean seawater; they are then teased with clean needles, which results in setting free a large number of ova and spermatozoa. The contents of the glass are well stirred up, and allowed to stand for two or three hours. The fragments of the Myzostoma are removed by the aid of needles, and what remains thrown into a

---

[1] A. Agassiz, Ann. Lyc. Nat. Hist., VIII., p. 318.
[2] Mitth. d. zool. Station z. Neapel, V., p. 545-547, Nov., 1884.

glass full of clean sea-water. This glass should be covered, and the water kept well aërated by means of an air-apparatus such as has been devised by Dr. Andres. As the ova and larvæ always remain at the bottom, the greater portion of the water can be poured off, and so renewed every second or third day.

It is well in each experiment to take at least four or five Myzostoma; for they are hermaphrodite, and it is therefore necessary to guard against self-fertilization as far as possible. The number of instances in which it can occur will, of course, be diminished by increasing the number of individuals in each experiment.

The post-larval stages may be obtained by slowly killing the Comatulæ in a mixture of sea-water with 10 % alcohol. When dead, each individual is shaken in the mixture by the aid of forceps, in order that the Myzostoma may fall off. The most of the water is next poured off, and alcohol gradually added until a strength of 90 % is reached. A large number of parasites may then be stained together, and mounted for microscopic examination.

THE EGGS OF SAGITTA.[1]

To obtain the eggs of Sagitta, it is only necessary to place freshly-captured animals with gravid reproductive organs in a small vessel for a few hours. The eggs will be found floating just under the surface of the water. Although the eggs are comparatively large, it requires favorable light to see them, as they are perfectly transparent. They can be fished up by means of a glass tube or a large pipette.

The earliest stages may be best obtained by cutting open gravid specimens, allowing the sexual products to flow out through the opening thus made. In this way, the process of fecundation and the production of polar globules may be witnessed.

They can be easily studied in a living condition, and the views thus obtained may be supplemented by treatment with 1–2 % acetic acid.

---

[1] O. Hertwig. Jenaische Zeitschrift, XIV., 1880, p. 271.

PREPARATION OF THE OVA OF THE CRAYFISH.[1]

A. Surface Preparations of the Early Stages.— 1. Heat the ova in a vessel partially filled with water, raising the temperature gradually to 80° C.

2. After cooling, make an incision in one side of the chorion with a sharp scalpel, and then remove the chorion with needles. The thin membrane inside the chorion may then be easily removed.

3. With a very sharp razor, wet with strong alcohol, while holding the ovum between the fingers, cut off the embryonic side of the ovum in the form of a thin, plano-convex disc.

4. Stain the disc in Ranvier's picro-carmine (about thirty minutes).

5. Wash, treat with alcohol, and mount in balsam.

B. Preparation for the Microtome.— 1. After heating in water to 80°, place the ovum (still in its chorion) for 24 hours in bichromate of potassium (1-2%) or in chromic acid ($\frac{1}{2}$%).

2. Wash for 24–48 hours in water, which must be often changed.

3. Harden in weak alcohol (60-70%) 24 hours, and for the same time in absolute alcohol.

4. Remove the chorion, and stain *in toto* in dilute Beale's carmine, or in carmine acetate.

5. After treating with alcohol, cut off the embryonic surface and imbed it in paraffine.

THE EGGS OF THE PHRYGANIDS.[2]

Time of Ovipositing.— The eggs of Neophalax are probably laid after dark. They were found in abundance from the middle of August till late in October. It was not ascertained how much earlier than the middle of August they may occur.

Place of Deposit.— According to Rathke, Zaddach, and

---

[1] H. Reichenbach. Zeitschr. f. wiss. Zool., XXIX., pp. 124–126, 1877.
[2] Wm. Patten. Quart. Jour. Mic. Sc., XXIV., Oct. 1884, pp. 549-558.

McLachlan, the Phryganids lay their eggs, arranged in a single layer, and enveloped in a transparent, gelatinous substance, on the under side of various water plants. The eggs of Neophalax, however, are deposited on the muddy bottom of slow running streams. They are found in greatest abundance in water four or five inches deep, and where there are a great many twigs lying on the bottom and projecting above the surface.

The Eggs and their Gelatinous Envelopes. — About 150 eggs are found in a single spherical, vesiculated mass of gelatinous substance. The eggs are greenish yellow, and their gelatinous envelopes have a strong odor of musk, which is perceptible even after they have been hardened by heat and preserved in alcohol for several months.

Preparation of the Eggs. 1. The best method of killing is to place the eggs in cold water and raise the temperature very gradually to about 80° C.; the proper time to check the rise in temperature is indicated by the eggs becoming hard and white.

2. After cooling, they are removed to 20% alcohol, which is increased by 10% once or twice a day until it has reached the required strength (about 96%).

3. Only alcoholic stains will penetrate the chorion, and of these Kleinenberg's hæmatoxylin and a 70% solution of cochineal gave by far the best results. Partial and varied success was obtained with safranin dissolved in 90% alcohol. The eggs were allowed to remain in hæmatoxylin 5–6 days, and were then decolorized very gradually in acidulated alcohol (1 drop of strong hydrochloric acid to 20 grams of alcohol), in which they remained several days. They were then transferred to pure alcohol, which was changed once or twice until the object regained its violet color.

Imbedding. — The eggs are clarified in benzole instead of clove oil, as the former penetrates almost instantly, and is less likely to cause shrinkage or to render the yolk hard and brittle.

After remaining 30 minutes in benzole, they are placed in a saturated solution of paraffine in benzole, and then in pure melted paraffine (melting at 58° C.).

Cutting. — The blocks of paraffine in which the objects are imbedded should have the form of a triangle, the base, which is the shortest side, being just wide enough to hold the object, and the altitude about five or six times the length of the base. The apex of the triangle is directed towards the edge of the knife, which, if placed at an angle of 10° or 12° to the direction in which it slides, will reach the apex of the triangle first, and cut a closely rolled shaving of paraffine. The section of the object, since it lies in the last portion of the section cut, will occupy the outside roll; and, if the triangle is long enough, will not cover more than one half the circumference of the roll. The whole roll is placed on the slide with the section of the object down. When the slide is warmed, the melting paraffine flattens the section, and thus the trouble of unrolling is avoided.

By this method, very thin sections (about one four-hundredth of a millimeter) may be obtained.

TREATMENT OF THE OVA AND EMBRYOS OF THE APHIDES.

Witlaczil[1] gives a lengthy paper on the development of the Aphides, and in it the following information on methods: —

The embryos of the viviparous Aphides were examined in a weak salt solution (1½ per cent), in which they live for about an hour. The ovaries contain embryos in different stages of development, and have to be isolated for study.

The early stages in the development of the ova may be studied to advantage after treatment with hydrochloric acid (3 %), or acetic acid, as these reagents partially dissolve the yolk elements and thus render the preparation more transparent. The later stages, on the contrary, are rendered more opaque by the same treatment.

The ovaries of viviparous Aphides were prepared for sectioning, by Ludwig Will,[2] in the following manner.

The Aphis is killed in water heated to about 70° C., then hardened in successively higher grades of alcohol. In order to color *in toto*, the cuticula must be punctured with a sharp

---

[1] *Zeitschr. f. wiss. Zool.*, XL, Heft 4, p. 563–564.
[2] Semper's Arbeiten, VI. 1883.

needle, and then the dye will penetrate easily to all parts. As very thin sections are required for the study of such small elements, it is best to use dyes that stain very deeply, such as borax-carmine and hæmatoxylin. It is further necessary to employ either the shellac or the collodion method of fixing the sections, otherwise important parts are liable to drop out of place.

THE EGGS OF LIMULUS.

The eggs of Limulus are laid between and above tide-marks about the first of June, on the south side of Cape Cod. It appears probable that the period of ovulation is rather prolonged, as I have found at Annisquam eggs well matured in the oviduct the last week in July. In hunting for the eggs no indications can be followed. The eggs are deposited in the sand and must be searched for at random. In general appearance, as they lie loosely in the sand, they resemble seed pearls; the general color is a dirty green or ashy gray, but pink eggs are common, and these develop as well as any. My attempts at artificial impregnation were apparently unsuccessful, though I learn from Dr. E. B. Wilson, that Mr. H. L. Osborn, of Johns Hopkins University, tried it and found that the early stages developed slowly, and possibly I did not keep the eggs long enough.

The eggs obtained from the sand possess great vitality and will live for months in confinement. Embryos from eggs obtained in June were kept in a small aquarium until November, the loss of water from evaporation being made good by daily additions of fresh water. In Dr. Lockwood's experience, eggs passed through the winter, hatching in the spring. I obtained my best results in the following ways: —

For surface views, a light staining with osmic acid ($\frac{1}{10}$ %) proved best, rendering visible details not otherwise readily seen. For sections, successive grades of alcohol (50, 75, 90 %, and absolute) were at least equal to chromic acid, Merkel's fluid, Perenyi's fluid or Müller's fluid. With chromic acid they show a tendency to shrink. Corrosive sublimate renders the yolk very hard and brittle, while the coagulation of the albumen by heat, so successfully employed in studying many arthropods, proved in my hands worthless. The greatest difficulty with the earlier stages was in freeing the egg from the chorion. In most cases I had to cut chorion and egg together, and the results were far from satisfactory. The eggs stain very slowly. My best preparations were obtained by over-staining the sections on the slide with eosin or hæmatoxylin, and then withdrawing the excess of color by acid alcohol.— J. S. KINGSLEY.

MOLLUSCAN EGGS.

For embryological study, Fol[1] recommends *Planorbis* as the most favorable object among fresh-water molluscs.

Planorbis contortus deposits in May and June; *P. marginatus*, from July to September.

Helix deposits its eggs in the earth and then covers them. If one marks the spots where they are seen preparing their nests and does not disturb them, the eggs may be found on the following day. The first stages of cleavage can only be obtained by killing specimens ready to deposit. The eggs, if carefully removed from the cocoon, and placed in a compressorium where they can be kept moist, and supplied with air, will develop normally for a week or more.

THE EGGS OF THE SLUG (*Limax campestris*).[2]

Place of Deposit. — The eggs are found in moist places, protected from the drying influences of direct sunlight and currents of air. They are frequently met with in the vicinity of a small stream, some stagnant pool, or in low meadow-land. In such localities, loose piles of decaying wood are often chosen for the deposit of eggs. At other times, when the ground is less protected, they may be found under loose stones, or even in the bed of some spring-time water channel, where crevices in clumps of earth afford protection. They are laid at all hours of the day, but, as a rule, early in the morning.

Mode of Culture. — The most convenient way of obtain-

---

[1] Arch. de Zool. expérimentale, VIII., p. 103.
[2] Mark. Bull. Mus. Comp. Zool., VI., Part 2, No. 12, 1881.

ing material is to keep the slugs in confinement. If properly cared for, they will thrive, and produce eggs, even during the winter. They may be kept in glass or earthen dishes covered with gauze or with perforated tin covers, and supplied with moss, which must be kept wet.

Food.—Lettuce, plantain leaves, cabbage, &c. During the winter they may be kept on wheat bran. To avoid the rapid growth of mould which follows when the bran is sprinkled over the wet moss, the bran may be kept in a watch-glass. If they are not kept supplied with food, they will devour one another.

Appearance of the Eggs.—The eggs are usually found in clusters of about a dozen, though the number may vary from one to thirty or forty. Sometimes they are loosely collected together, or even scattered over an area of a few inches; at other times, they are closely packed in more or less rounded masses. They are often arranged in rows in the narrow cracks of moist, decaying wood, or in the chinks in caked earth. They are frequently united in the form of a rosary.

Egg-Envelopes.—The ovum proper is just visible to the unaided eye as a whitish dot, having not more than one twentieth the diameter of the whole egg. The ovum is imbedded in a yellowish, viscid substance, called the albuminous envelope. This central mass, occupying from $\frac{2}{3}$ to $\frac{4}{5}$ the diameter of the egg, is invested by a very firm, structureless membrane (*membrana albuminis*). External to this are two thick, transparent envelopes, the inner of which is semi-fluid and homogeneous, while the outer is laminated and elastic.

Preparation of the Ova.—1. Treat with acetic acid (1-2 %) four hours or more.

2. Remove the outer envelopes in the following manner:—

Hold the egg between the thumb and finger of the left hand, thrust the point of fine scissors through the two external envelopes, and then make a cut sufficient to allow the resistant membrane (*membrana albuminis*) with its contents to escape. Place the egg in a drop of water on a slide, and remove the external envelope. Portions of the semi-fluid envelope will remain attached to the surface of the resistant membrane. The next step is more difficult. Thrust a sharp needle, under the dissecting microscope, *quite through* the resistant membrane, and thus hold the egg, while, with another needle, a second puncture, followed by a gash (by gently pulling in opposite directions with the needles), is made, which will allow the ovum to escape with the albumen. Float the ovum, by the aid of one needle, into a fresh drop of water near one end of the slide, strand it, and quickly wipe away (with a cloth wound over the tip of the fore finger) all remaining albumen and water. Add a drop of fresh water, and float the ovum back to the centre of the slide. Arrange a cover-glass with wax feet, and adjust it under the microscope.

3. Allow a drop of Beale's carmine to flow under the cover, and examine at once. The ovum may remain from 1 to 2 hours in the staining fluid.

4. Draw away the carmine with the aid of a brush, and at the same time allow dilute glycerine to replace it.

5. Replace the dilute with pure glycerine.

This method will serve for studying the earlier changes in the ovum. In later stages osmic acid works better than acetic acid, as it hardens the albuminous envelope as well as the ovum.

EGGS OF THE STARFISH.

The following extract is from A. Agassiz's Embryology of the Starfish [1]: —

Sexual Distinction. — The males and females of our common species of Starfishes, *Asterias vulgaris* and *A. arenicola*, can readily be distinguished by their difference in coloring, all those having a bluish tint being invariably females, a reddish or reddish-brown color indicating a male. When cut open, so as to expose the genital organs, the difference between the males and females is still more striking. The long, grape-like clusters of reproductive organs, extending from the angle of the arms, on both sides of the ambulacral system, to the extremity of the rays, present very marked differences in the two sexes. The ovaries are bright orange, while the spermaries are of a dull cream-color. At the time of spawning, which is very different in the two species mentioned above, the genital organs are dis-

---

[1] Mem. Mus. Comp. Zool., Part First, pp. 3-5.

tended to the utmost, filling completely the whole of the cavity of the ray, the abactinal system itself being greatly expanded by the extraordinary development of these organs.

Artificial Fecundation. — If we take a male and female starfish in this state, and cut a portion of the genital organs into small pieces, we shall find that the eggs and spermatozoa escape in such quantities as to render turbid the water in which they are placed. Throwing these small pieces of the genital organs into shallow dishes containing fresh sea-water, and stirring the mixture thoroughly to ensure the contact between the spermatozoa and the eggs, will be sufficient to fecundate the latter. In order to make the operation perfectly successful, some precautions are necessary: all the pieces of the genital organs which are left after repeated stirring must be carefully removed; there must not be too many eggs in one dish, so that the water can have free access to them in every direction. The removal of the remnants of the ovaries and spermaries is very necessary, as the pieces which remain clotted together decompose very rapidly, and endanger the safety of the eggs, even when the water can be changed with the greatest facility. As soon as the fecundation is fulfilled, the water in the dishes must be repeatedly changed, until it becomes perfectly clear; for the presence of too many spermatozoa, rendering the water milky, prevents a favorable result. It is best only to use one male and one female for the mixture in each vessel, as eggs taken from many individuals lessen the chances of success. The eggs sink to the bottom, so that the water can be poured off and changed without much danger of throwing them away. Immediately after the mixture is made, the water should be changed three or four times in succession; after that, every half hour until the fourth hour, when an interval of two to four hours may elapse before renewing the water. As it is extremely difficult to change the water after the embryos have hatched and are swimming freely about in the jar, without losing many of them, it is advisable before they hatch, which is about ten hours after the fecundation, to reduce the water to a minimum volume, and then simply to add a little fresh sea-water and remove the contents of the vessel to larger and larger jars. In this way the water can be maintained sufficiently pure until the young embryos have taken the habit of swimming near the surface, when it may all be drawn off by means of a siphon. A great deal of time and trouble will be saved by this mode of procedure, and fewer specimens lost. The jars containing the eggs should be kept in a cool place; the most convenient method of securing a low and even temperature is to place the small jars in large tubs filled with cold water.

BALANOGLOSSUS.

The specimens **first discovered were found at Charleston, S. C., by Stimpson, over thirty years ago, and described by** Girard, under the name of *Stimpsonia aurantiaca*.[1] Mr. Agassiz has found it at Beverly, Mass., and at **Newport, R. I.**

Packard collected it at Beaufort, N. C.; and Verrill reports it from Fort Macon and Naushon **Island.**

It is found at low-water mark, buried **in the sand to a depth** of eight to fourteen inches. Its burrowings **are recognized by** peculiar elliptical coils of **sand** which are thrown out at the surface.[2] The burrow **is lined by a "thick mucous layer,"** which forms a sort of case **in which the worm is able to move** up and **down.**

The earliest free-swimming larvæ **are found,** according to Mr. Agassiz, about the first of August, **at Newport, and are** captured by surface skimming.

According to Bateson,[3] **who has recently studied the species** occurring at Hampton, Va., the sexual **products escape by a** rupture of the body-wall **in both sexes. The** eggs are small, ovoid, opaque, grayish yellow in color, and about $\frac{3}{_{}}^{mm}$ in length; they are enclosed each in an elastic, transparent membrane. The spermatozoa dehisce in lobate spermatophoric masses.

Bateson[4] made use of the following method for obtaining the larvæ and embryos: —

1. The mud inhabited by Balanoglossus is placed in a glass

---

[1] Verrill. Report 1871–1872, p. 351-2.
[2] A. Agassiz. Mem. Am. Acad. Arts and Sciences, IX., p. 421, 1873.
[3] Quart. Jour. Mic. Sci., XXIV., No. 94, p. 209, 1884.
[4] Quart. Journ. Micr. Sc., XXV. (Supplement, 1885), pp. 115, 116.

vessel of water, and stirred up, avoiding rotatory movement, until the whole is in suspension.

2. A number of adult individuals cut into very fine pieces are next thrown in, and the whole left to settle for a few minutes. These pieces sink to the same depth as the larvæ, and thus serve to mark the important layer. Practice enables one to dispense with this step, the right layer being recognizable by the size and character of the particles composing it.

3. The water and lighter particles in suspension are then siphoned off until the layer of chopped Balanoglossus is reached. This layer is drawn off, and examined separately. In this manner all the animals living in several hundredweight of mud may, in an hour or two, be collected into about a pint of water, and sorted with a simple microscope. This was generally done by rotating a little of the water in a shallow saucer with a slight peripheral groove. The larvæ then all lie in the groove, which may be passed under the lens by rotation.

Preservation. — The larvæ are placed for less than a minute in —

Corrosive sublimate (sat. sol.) . . . . 2 parts
Glacial acetic acid . . . . 1 "

After washing with water, they are passed through 30 %, 50 %, 70 %, and 90 % alcohol. While this method gave the best results, on the whole, the softer parts were found to be best preserved in specimens that were treated with Perenyi's fluid one hour, and then transferred to 90 % alcohol.

THE EGGS OF DOLIOLUM.[1]

The ripe egg of Doliolum passes through the genital orifice into the atrium or cloacal chamber, and is soon carried out of this chamber with the water. The egg does not swim on the surface of the water, but sinks to the bottom and there remains until the larval tail is formed. There is no way, therefore, of obtaining material for the study of the development except

---

[1] B. Uljanin. 'Fauna u. Flora des Golfes von Neapel.' X. Monographie: Doliolum, pp. 46-49. 1884.

that of keeping the two sexes in confinement. Uljanin kept them in glass vessels filled with filtered sea-water and covered with a glass plate. The vessels were placed in tanks, and a small stream of water allowed to flow steadily on to the glass cover. The evaporation of the water from the sides of the vessel kept it at a low and even temperature, which was found to be favorable for keeping the animals and the eggs alive.

The fresh-laid egg is inclosed in a follicle, consisting of a single layer of cells. The fecundation and the metamorphosis of the germinal vesicle take place after the egg has escaped from the cloacal chamber. The development is remarkably rapid, the tailed larva appearing in about 6 hours after fecundation of the egg.

PELAGIC FISH EGGS.

The transparent eggs of various Teleostei found floating on the surface of the sea during the summer months are extremely favorable objects for the study of vertebrate development, notwithstanding the difficulties in the way of hardening.

These eggs are found at Newport all through June, July, August, and September, and probably in April and May. They occur in greatest abundance in June, falling off considerably by the middle of July, and becoming less and less abundant from this time onward.

The eggs are readily collected by surface skimming with a fine hand net or a tow net. The calm streaks formed by tidal eddies are the best collecting ground.

To find the eggs, after they have been collected, the water containing them must be poured into glass vessels and examined on a black ground. Mr. Agassiz employs glass pans, and has one end of the table covered with black tiles (the rest with white tiles), which makes a very convenient ground for purposes of this kind.

Various species of eggs will of course be found together, and the distinguishing marks are often more or less difficult to recognize. But the difficulties in the way of *identifying* species are incomparably greater than those of "sorting" the eggs.

The most important guides to identification have been thus stated by Mr. Agassiz: —

"The pigment spots of the surface of the yolk, and those characteristic of different species of fish embryos, begin to make their appearance at very different times in the many species we have examined. Hence, until the characteristic pigment pattern of an embryo is pretty well known, it is easy to confound the eggs of very different species. The position and shape of the otoliths and the degree of development of the pectorals become also excellent guides to the identification of eggs well advanced in their development. The differences in the young embryos on hatching are very considerable, and in these earlier stages the degree of development of the head, the proportional size of the yolk-bag, the shape of the embryonic fin, the position of the vent, and the pattern of the pigment spots, are all of great use in the identification of the species."

"The presence or absence of an oil globule is an excellent guide in the identification of the egg; the size of this globule is, however, quite variable. In one of the species of Cottus there are many globules present, and the number of these varies from sixteen to thirty-two for this species. In another species, (Hemitripterus) in which there is generally only one globule, it is not an uncommon occurrence to find two globules."

The size of the egg varies, as a rule, within narrow limits ($.01-.05^{mm}$), and is therefore an important aid in identification.

In studying the living egg, profile views and optical sections are obtained by simply tilting the microscope, the tube being inclined at different angles between the vertical and horizontal positions, according to the view desired. The egg should be confined in a live-box, in a drop of water, without pressure. Thus confined, the egg will live till the time of hatching without change of water, and its position is under perfect control.

I have experimented with all the hardening reagents in common use, and have failed to find any completely satisfactory method of preserving the vitellus. Even the germinal disc cannot be well preserved by any of the ordinary hardening fluids. Kleinenberg's picro-sulphuric acid, for instance, causes the cleavage products to swell, and in many cases to become completely disorganized. The embryonic stages can be hardened in chromic acid (1%), but the yolk contracts considerably without becoming well hardened. The best preparations of the cleavage stages have been obtained with osmic acid, followed by a modified form of Merkel's fluid. This fluid consists of chromic acid ($\frac{1}{4}$%) and platinum chloride ($\frac{1}{4}$%) mixed in equal parts. Thus prepared, it causes maceration of the embryonic portion of the egg. By using a stronger chromic acid (1%) and combining it as before with the same volume of platinum chloride ($\frac{1}{4}$%), everything may be well preserved and hardened except the yolk. But this fluid cannot be used with success unless the egg has been first killed by another agent; for eggs placed in this fluid continue to live for a considerable time, and may even pass through one or two stages of cleavage. It is therefore necessary to use some reagent that kills instantly. For this purpose a weak solution of osmic acid may be used.

The eggs are placed in a watch-glass with a few drops of sea-water, and then a quantity of osmic acid ($\frac{1}{2}$%) equal to that of the sea-water is added. After five to ten minutes the eggs are transferred to the mixture of chromic acid and platinum chloride, and left for twenty-four hours or more. This fluid not only arrests the process of blackening, but actually bleaches the egg to a considerable extent. After this treatment it is an easy matter to separate the blastoderm from the yolk by needles; and the preparations thus obtained may be stained at once, and then treated with alcohol and mounted in balsam.

The value of such preparations must be measured by the accuracy and clearness with which they present the conditions in the living state. A careful study of the preparations shows that the method can be relied on in every particular. The osmic acid fixes the living conditions more perfectly than any other reagent at present known, and the chrom-platinum mixture completes the work of hardening without shrinkage or swelling. The finest details of the cleavage lines, the cleavage cavity, and the nuclear figures are well preserved. The relation of the blastoderm to the protoplasmic mantle enveloping the vitellus, and all the particulars in regard to the origin of the so-called "free nuclei," are satisfactorily shown. No violence is required in order to free the blastoderm from the yolk, as a clean separation is usually effected by the action of the acids. In this separation the protoplasmic mantle invariably goes with the blastoderm, in the older as well as in the younger stages of development.

For sectioning, the embryonic portions of **the** egg need not **be** separated from the yolk. But before transferring the eggs **from the** chrom-platinum solution to the different grades of **alcohol (50 to** 100%), **the** egg-membrane should be broken or **perforated by the aid** of needles **on** the side opposite the blastoderm, **in order that the alcohol** may reach the egg readily, as **otherwise the membrane wrinkles** badly, and often injures the **embryonic portion.** For the embryonic stages, the above **method of hardening has not been** altogether satisfactory. **The embryos are brittle, and the boundary lines** between the **different parts are not always sufficiently clear.** A few of **these stages were hardened in Perenyi's fluid,** and the sections **have proved much more instructive than any** obtained from eggs hardened in other fluids.

Perenyi **recommends leaving eggs from four** to five hours in **this fluid. Two hours' immersion is** certainly sufficient for **pelagic fish** eggs, and **probably** even a shorter time would **do. Among the methods of** staining recommended by Perenyi may **be mentioned that of** mixing borax-carmine or picro-carmine **with the** hardening fluid. The addition of the staining **fluid** produces a precipitate which should be removed by filtering. The filtered mixture both hardens and stains at the same time, which is certainly an advantage. After an immersion of a few hours in this fluid, the eggs should be transferred first to 50% alcohol (5 hours), then to successively higher grades. This method of hardening does not render the eggs brittle.

EGGS OF THE SALMONIDÆ.

The following method is recommended by Henneguy[1] for the eggs of Salmonoids:—

---

[1] Bull. de la Soc. Philomathique, No. 1, 1879, pp. 75-77.

1. Osmic acid (1%) for a few minutes till the eggs show a light brown color.

2. Transferred to Muller's fluid and left for a few days.

The egg-membrane is cut with a fine pair of scissors while in this fluid. The yolk dissolves, leaving the hardened embryo and the cortical layer (periblast) free, so that they can be easily removed and prepared for sectioning or for mounting *in toto*.

4. Stained in picro-carminate of ammonia.

5. Dehydrated with alcohol of 80% followed by absolute alcohol.

6. Immersed in collodion 24 hours.

7. Placed on a piece of elder-pith soaked with alcohol, and covered with a layer of collodion.

8. Sectioned as soon as the collodion has hardened sufficiently, then mounted in glycerine.

Trutta fario.[1] — 1. Chromic acid ($\frac{1}{2}$%) 24 hours.

2. Distilled water, 2 hours. The egg membrane expands, and may now be easily removed.

3. Washed in distilled water 12 hours.

4. Absolute alcohol, glycerine, and aq. dest. in equal parts, 4 hours.

5. Absolute alcohol.

6. Böhm's carmine acetate, 1-2 days.

7. Mixture of water (50 vols.), glycerine (50 vols.), and muriatic acid ($\frac{1}{2}$ vol.), for a few minutes.

8. Washed in water 4-5 hours.

9. Absolute alcohol 12 hours, preparatory to imbedding in paraffine.

THE EGGS OF THE AMPHIBIA.

Our common frogs and toads begin to spawn soon after the first warm days of spring. The eggs of *Rana* are generally found first, and the spawning season lasts for about a month, beginning (for this climate) the first or second week in April. The spawning season begins about the same time with *Amblystoma*, but not until two or three weeks later with the bull-frog (*Rana catesbeiana*).

---

[1] Kupffer. His and Braune's Archiv, **Anat.** Abth., 1882.

The eggs of the frog cohere in a mass, and only a few minutes are required for their deposit; while those of the toad are excluded in a long string of jelly, the process lasting for hours. The eggs of *Amblystoma* are laid in gelatinous masses a little smaller than that of the frog. The eggs are light brown, and thus easily distinguished from the dark-brown eggs of the frog or the almost black eggs of the toad.

Our common newt, *Diemictylus miniatus*, deposits its eggs singly on aquatic plants. The whitish ectosac is elliptical, and much harder and tougher than the envelope of the frog's egg. The egg is pale brown or flesh-colored, and, as it is closely wrapped up in leaves, which are glued together by its adhesive coating, it is not easily found. If specimens are captured at the beginning of the spawning season, and the pairs found copulating are isolated in vessels containing only a few water-plants, the eggs are readily obtained.

The process of copulation, in which the male clasps the female around the neck with his hind legs, generally lasts from one to two hours. I have never seen eggs extruded at the time of copulation, and am inclined to think, from what I have seen, that fecundation takes place before deposit. The spermatozoa would easily find their way to the cloaca of the female through the fanning movement kept up by the tail of the male; but it is more difficult to account for their actual entrance, as there is no contact of the labia of the cloaca in the two sexes.

Preparation for Sectioning. — I have obtained good results with the frog's eggs by the following method: —

1. Small pieces of the gelatinous mass — each containing from 5 to 15 eggs — are cut free with scissors, and placed in $\frac{1}{4}$ % osmic acid for 20-25 minutes.

2. Transferred directly to the chrom-platinum solution recommended for fish eggs (24 hours).

3. The eggs are next placed in water, and freed from their gelatinous envelopes by the aid of sharp needles and a dissecting microscope. The removal of the envelopes is accomplished without much difficulty after a little experience.

4. Wash in flowing water, or in water often renewed, for at least two hours.

5. Treat with 50 % alcohol (1-2 hours), 70 % (5-10 hours), then preserve in 80 % or 90 %.

If sectioned soon after hardening and dehydration with alcohol, the eggs will cut without crumbling when imbedded in paraffine. They may be stained, after passing through 90 % alcohol, with borax-carmine; but the staining will be of little or no help in the earlier stages.

HERTWIG'S METHOD OF PREPARING AND CUTTING AMPHIBIAN EGGS.[1]

Although the amphibian egg has long been a favorite object of study among embryologists, — and quite as much so since section-cutting came into vogue as before, — comparatively little progress has been made in overcoming the difficulties that attend its preparation for the microtome. The chief difficulties are found in freeing the egg from its gelatinous envelope, and in preparing it so as to avoid brittleness.

The best method that has thus far been proposed for these eggs is unquestionably that of O. Hertwig, and I shall therefore give it in detail.

1. In order to facilitate the removal of the gelatinous envelope, the eggs are placed in water heated almost to boiling (90-96° C.) for 5-10 minutes. The eggs are thus coagulated and somewhat hardened, while the envelope separates a little from the surface of the egg, and becomes more brittle. The envelope is then cut under water with sharp scissors, and the egg shaken out through the rupture. With a little experience a single cut suffices to free the egg.

2. By the aid of a glass tube, the egg is taken up and transferred to chromic acid ($\frac{1}{2}$ %), or to alcohol of 70, 80, and 90 %. Chromic acid renders the egg brittle, and the more so, the longer it acts; therefore the eggs should not be allowed to remain in it more than twelve hours. While eggs hardened in chromic acid never change their form or become soft when transferred to water, those hardened in alcohol, when placed in water or very dilute alcohol, lose their hardness, swell up, and often suffer changes in form.

---

[1] Jenaische Zeitschrift für Naturwissenschaft, XVI., p. 249, 1882.

3. Alcoholic preparations are easily stained; but chromic acid preparations are stained with such difficulty and so imperfectly that Hertwig omitted it altogether.

There is an important difference between alcohol and chromic acid in their effect on the pigment of the egg. Chromic acid destroys the pigment to some extent, and thus obliterates, or at least diminishes, the contrast between pigmented and non-pigmented cell-layers. As the distribution of the pigment is of considerable importance in the study of the germ-lamellæ, it is well to supplement preparations in chromic acid with those in alcohol, in which the pigment remains undisturbed.

4. Eggs hardened in chromic acid were imbedded almost exclusively in the egg-mass recommended by Calberla. The great advantage offered by this mass is, that it supplies a sort of antidote to the brittleness of the egg. It glues the cell-layers together, so that the thinnest sections can be obtained without danger of breaking.

5. As the dorsal and ventral surfaces and the fore and hind ends can be recognized in very early stages, it is important to know precisely how the egg lies in the egg-mass, in order to determine the plane of section. In order to fix the egg in any given position in the imbedding mass, Hertwig proceeds as follows:—

a. A small block of the hardened mass is washed in water to remove the alcohol; and in the upper surface of the block, which has been freed from water by the aid of filtering paper, a small hollow is made. This hollow is then wet with the freshly prepared *fluid* mass.

b. The egg is washed in water, to remove the alcohol, placed on a piece of filtering paper to get rid of the water, turned on the paper by a fine hair brush, until it has the position desired; the point of the brush is next moistened and pressed gently on the upper surface of the egg; the egg adheres to the brush, and may thus be transported to the hollow prepared for it in the block.

c. After the egg has thus been placed in position, a drop of absolute alcohol, carefully applied, will coagulate the "fluid mass" with which the hollow was wet, and thus fix the egg to the block. The block is again washed, and finally imbedded in the egg-mass (cf pp. 106-107).

THE INFLUENCE OF GRAVITATION ON THE DIRECTION OF CLEAVAGE-PLANES.

The slimy envelope of the frog's egg undergoes peculiar changes the moment it comes in contact with water. It swells up very quickly, assumes a gelatinous consistency, and loses its adhesive property. The cavity of the envelope thus becomes somewhat larger than the egg and fills with water, in which the egg can rotate easily. Under these conditions, the egg takes its normal position, with its pigmented pole uppermost, very soon after deposit. In this position the axis of the egg is vertical, and the first cleavage-plane is meridian, coinciding with the axis. It has generally been supposed that the direction of the first plane of cleavage is determined by some necessary relation to the axis of the egg; but the means of testing the validity of this view is a recent discovery, for which we are indebted to E. Pflüger.[1] A series of brilliant experiments was undertaken by this author in the hope of settling the question, whether the direction of the first cleavage-plane is determined by the direction of the axis of the egg, or by the direct influence of gravitation. These experiments were made possible by the discovery of a method by which the egg could be prevented from rotating and confined in any desired position. This was accomplished in the following manner: —

Pflüger. — Mature eggs were taken from the uteri and placed in *dry* watch-glasses (a single egg in each glass); then a drop of water containing spermatozoa was added, and, after a few seconds, drained off. Thus treated, the mucous envelope does not swell sufficiently to allow the egg to rotate; but it adheres to the glass, and, at the same time, presses upon the egg just enough to hold it in position. If the eggs are protected against drying by keeping in a moist chamber, development appears to go on as well as under normal conditions.

---

[1] Archiv. f. d. ges. Physiologie, XXXI., p. 312, 1883.

The experimenter finds thus a very ready means of giving the axis of the egg any desired inclination, and is in a favorable position for studying the effect on the direction of the cleavage plans.

Born.—Born,[1] who has followed up the line of investigation started by Pflüger, gives some valuable details of the method of procedure. A number of shallow, round, flat-bottomed glasses of different sizes served for receiving the eggs. These were provided with glass covers, which were so adjusted as to admit air without allowing much evaporation.

The eggs were fixed on round glass plates, somewhat smaller than the flat bottoms of the dishes on which they rested. In one diameter of each plate was etched an arrow, and on each side of the arrow four or five eggs were placed, and their position noted in a diagram. In addition to this, two diagrammatic sketches of each egg were made on two rectangular pieces of paper, one to show the upper, the other the lower aspect of the egg, at the moment it was fixed on the plate. Each egg was further designated with its number and time. In order to view the eggs from below, the plate was simply inverted over a smaller glass dish containing a little water. A damp chamber was thus formed, so that the eggs were not exposed to evaporation during the preparation of the diagrams. The sketches were made for the most part without the aid of the camera lucida and microscope.

Fecundation was accomplished by adding, with a brush, to each egg a small drop of water containing spermatozoa. If too large a quantity of spermatic fluid was added, the excess was removed by the aid of a bit of blotting-paper.

Born remarks that most of the eggs in his first experiments were lost by undue exposure to evaporation; and adds that danger in this direction may be guarded against by covering the plate which carries the eggs with a fine spray of water, as often as it is exposed for the purpose of making drawings.

Eggs were prepared for section-cutting in the following manner:—

---

[1] Archiv f. mik. Anat., XXIV., H. 4, pp. 476–480, 1885.

1. For orientation, the gelatinous envelopes were marked with red and green size-colors, which were liquidized by heating. The melted coloring substance was applied in drops with a very fine brush in such a manner as to indicate the position of the two poles of the egg. The diagrams were marked with the same colors at corresponding points, in order to render identification easy.

2. After marking, the eggs were killed by placing them (together with the glass plate) in oil heated to about 90° C., then transferred to 75% alcohol, and, after a few hours, to 80% alcohol. They were next separated from the glass plate by means of a thin knife pressed against the plate, and then left in 80% or 90% alcohol for preservation.

3. The eggs were imbedded in paraffine in the usual way, the color-marks, still visible on the envelopes, serving for orientation.

4. Eggs thus treated were brittle, and complete sections could only be obtained in one of two ways — either by passing a hot spatula quickly over the cut surface of the paraffine block, or by covering this surface with a drop of melted paraffine, before drawing the knife each time. Painting with thin collodion would undoubtedly have served equally well, had the necessary precautions been observed.

The startling conclusion reached by Pflüger, that gravitation determines the direction of cleavage-planes, opened at once a new field of investigation, and called out other important methods of testing this question.

Roux. — Roux[1] placed frog's eggs, immediately after fecundation, in moist cotton in a centrifugal apparatus, the turning of which was sufficiently rapid to overcome the directive influence of gravitation, so that the axes of the eggs, instead of taking a vertical direction, remained in whatever position they chanced to be at the start. The result was that the first cleavage-plane invariably coincided with the axis of the egg, and that the development was in all respects the same as under normal conditions.

---

[1] Breslauer ärztlichen Zeitschrift, No. 6, 1884.

Rauber. — Rauber[1] endeavored to confine the trout's egg in abnormal positions by means of peculiarly constructed silver-plated clamps, but was unsuccessful. In the centrifugal apparatus the axis of the trout's egg took a horizontal position, with the germinal disc directed centripetally, and the yolk centrifugally. In this position, the eggs developed normally, from which Rauber inferred that gravitation could be replaced by centrifugal force, but that a directive force of some kind must, in all cases, be present, under the influence of which development takes place.

O. Hertwig. — The small, transparent eggs of the sea-urchin are perfectly spherical and free from food-yolk, and in them the centre of gravity may be assumed to coincide with the geometrical centre. In these respects they were regarded by O. Hertwig[2] as favorable objects for studying the main question raised by Pflüger's investigations. For examination, the fecundated eggs were placed in a hanging drop of water on the under side of a cover-glass. The first cleavage-plane exhibited inclinations in every direction, varying between the vertical and horizontal position, from which it was inferred that its direction was not controlled by gravitation.

REPTILIAN OVA.[3]

1. The ova taken from the oviduct are opened in a dilute solution of osmic acid (1 %), and then the white removed as far as possible.

2. The osmic acid is then turned off, and a weak solution of chromic acid ($\frac{1}{4}$ %) added. 24 hours.

3. With a sharp, fine pair of scissors, cut around the germinal area, just outside its margin; and, after it has been completely encircled with the incision, float it carefully off from the body of the yolk.

4. The yolk and acid are next removed, and a copious supply of clean water added, which must be several times renewed.

---

[1] Berichte der naturforschenden Gesellschaft zu Leipzig, Feb. 12, 1884.
[2] Jenaische Zeitschrift, XVIII., pp. 178-182, 1885.
[3] Kupffer. His and Braune's Archiv, Anat. Abth., 1882, p. 4.

5. Calberla's fluid (glycerine, water, and absolute alcohol in equal parts) 3 hours.

6. Hardened in 90 % alcohol.

7. Stained in Böhm's *carmine acetate* 24 hours.

THE HEN'S EGG.

The very complete instructions given by Balfour[1] for studying the development of the chick may be assumed to be in the hands of every student of avian embryology, and need not, therefore, be summarized here. A very important addition to this branch of technique has recently been made by Duval,[2] whose methods I propose to give in detail. First in order and importance are the means and method of orientation. After the appearance of the primitive streak, at about the twelfth hour of incubation, it becomes easy to distinguish anterior, posterior, and lateral regions in the blastoderm. Hitherto it has been a matter of conjecture whether anterior and posterior regions became morphologically defined at any considerable time before the formation of this streak; and no one, before Duval, attempted to clear up the question, simply because it appeared impossible to find any means of orienting sections at an earlier date. Duval addressed himself to the task of finding out the transformations of the blastoderm, which lead up to the establishment of the primitive streak, and to this end he was compelled to seek, first of all, for some reliable means of exact orientation.

Method of Orientation. — It was noticed by Balfour, and confirmed by Kölliker, that the axis of the chick embryo lies constantly at right angles to the longer axis of the egg. If an egg, after one or two days' incubation be opened, while held in such a position that its large end is turned to the left and its small end to the right of the operator, it will be found that the caudal end of the embryo is directed towards the operator, while the cephalic end is turned in the opposite direction, as indicated in Fig. 31. Out of 166 cases Duval found only three that

---

[1] Appendix to the "Elements of Embryology."
[2] Ann. des sc. nat., XVIII., Nos. 1, 2, 3, 1884.

could be regarded as exceptions to the rule. Assuming that the orientation is the same before the appearance of the primitive streak, we have then a very reliable means of recognizing, even in the fresh-laid egg, when the blastoderm has a homogeneous aspect, the future anterior and the future posterior region. But this fact alone is not all that is required for complete orientation: the blastoderm must be hardened, and the means of orientation must be preserved. That portion of the vitelline sphere which bears the blastoderm

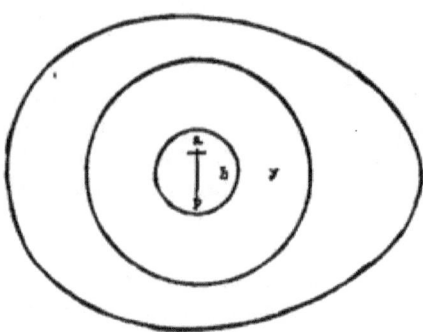

Fig. 31. — Diagram of hen's egg. *y*, yolk; *b*, blastoderm; *a*, anterior; *p*, posterior end of embryo.

must be so marked that the anterior and posterior regions of the blastoderm may be recognized after the process of hardening, and after the blastoderm, together with some of the circumjacent yolk, has been cut free from the rest of the egg. This may be done in different ways, according to the method employed in hardening.

I. Osmic-Acid Method. — 1. Make a triangular box without bottom, like that seen in Fig. 32, by folding a strip of paper 5^{mm} wide and 50^{mm} long.

Fig. 32.

2. After opening the egg carefully from the *upper* side, remove with a pipette the thin layer of albumen which lies above the cicatricula, so far as this can be done with safety.

3. Place the triangular box over the blastoderm in such a manner that the base corresponds to the future anterior region, and the apex to the future posterior region. While pressing slightly on the box in order to bring it into close contact with the surface of the yolk, fill it by means of a pipette with osmic acid ($\frac{1}{4}-\frac{1}{2}\%$), and allow the acid to act for some minutes.

4. As soon as the area inclosed by the box begins to blacken,

the whole should be immersed in a vessel of chromic acid, in which the paper box may be detached, and the vitelline sphere freed from the albumen and the shell.

5. The vitellus may now be transferred, by the aid of a very deep watch-glass, to another vessel of chromic acid, where it is allowed to remain one or two days, until the peripheral layers harden and form a sort of shell around the central portion which is still soft.

6. A triangular piece of this shell, inclosing the triangular area browned by the osmic acid, is next to be cut out with a pair of sharp scissors. The excised piece is then left a day or more in the chromic solution before treatment with alcohol.

II. Alcohol Method. — 1. Open the egg as before, and, without attempting to remove the albumen, place the triangular paper box over the blastoderm; slight pressure causes the box to sink into the albumen till it is brought into contact with the yolk. By the aid of a pipette, fill the box with absolute alcohol; this coagulates rapidly the inclosed albumen, while the albumen outside the box remains fluid.

2. After cutting the chalazeæ close to the vitellus, the fluid portion of the albumen is carefully drained off, leaving only the vitellus and the box with its coagulated contents in the shell.

3. The shell may now be filled with absolute alcohol until the yolk is completely covered, and then left for some hours, during which the more superficial layers of the yolk harden sufficiently to form a shell-like envelope of the softer central portion.

4. The triangular mass formed by the box and the hardened albumen is now ready to be cut out, in the same manner as in the osmic acid method. During this process, the paper box may become detached, either spontaneously or with some assistance; or it may adhere so firmly that it cannot be safely removed. There is no inconvenience in leaving it in place, as it will cut easily when the piece is ultimately sectioned.

5. The piece is further hardened 24 hours in absolute alcohol, then preserved in alcohol of $36°$ (80%).

III. Hot Chromic Acid. — 1. Treat with osmic acid as in No. I.

2. Place the whole in a solution of chromic acid, and heat to the point of boiling over a water-bath.

3. After cooling, cut out the triangular piece as in No. I. (6), leave it for a few days in chromic acid, then transfer to alcohol.

Imbedding and Cutting. — Duval imbeds, after each of the foregoing methods, in collodion (see p. 170). The surface of each section is collodionized some seconds before drawing the knife, by allowing a drop or two of thin collodion to flow over it.[1]

Staining. — The sections are placed in serial order on a slide, and then covered with picro-carmine strongly diluted with glycerine. The sections may be left in the staining fluid 24–48 hours, the admixture of glycerine preventing drying. After they are sufficiently colored, the staining fluid is allowed to drain off, and the slide is carefully washed with a pipette. The sections, still in place, are treated with successive grades of alcohol, and then mounted in balsam after being clarified in benzine ("benzine Collas").

Mounting the Blastoderm in toto. — During the first three or four days of incubation I have obtained good surface preparations of the blastoderm in the following manner: —

1. Break the shell by a sharp rap of the scissors at the broad end; then carefully cut away the shell, beginning at the place of fracture and working over the upper third or half.

2. After removing as much of the white as possible without injury to the blastoderm, place the rest of the egg, while still in the shell, in a dish of nitric acid (10%), deep enough to cover it.

3. The coagulated white should next be removed from the blastoderm by the aid of a brush or a feather, and the egg then allowed to remain in the acid 30 minutes.

---

[1] This method was recommended by Duval early in 1880. "De quelques perfectionnements à l'emploi du collodion en technique histologique." Société de Biologie, 1880. Later in same year came a similar account from Mason (Nov. 1880).

4. Cut around the blastoderm with a sharp-pointed pair of scissors, taking care to cut quickly and steadily. After carrying the incision completely round, float the blastoderm into a watch-glass, keeping it right side up and flat.

5. Remove the vitelline membrane by the aid of dissecting forceps, and the yolk by gently shaking the watch-glass and by occasional use of a needle. The yolk can sometimes best be washed off by means of a pipette.

6. Wash in water (several times changed).

7. Color deeply with carmine or hæmatoxylin.

8. Remove excess of color by soaking a few minutes in a mixture of water and glycerine in equal parts, to which a few drops (about 1%) of hydrochloric acid have been added.

9. Wash and treat 30 minutes with a mixture of

Alcohol (70%)	2 parts
Water	1 "
Glycerine	1 "

10. Transfer to pure 70% alcohol, then to absolute alcohol. Clarify with creosote or clove oil and mount in balsam.

The above method of treatment will also serve for blastoderms which are to be sectioned.

THE EGG OF THE RABBIT.[1]

The period of gestation for the rabbit is four weeks, and impregnation takes place immediately after littering. At the end of pregnancy, the ovary presents mature follicles containing eggs that are ready to be fecundated. The marks of maturity in the egg may be learned by comparing such eggs with eggs taken from the ovary at different times during gestation, and noting the differences. If a rabbit is killed two days before littering, eggs will be found which would have attained their complete development two days later. It is not, then, a difficult task to analyze the series of successive modifications which the egg undergoes during the four weeks of gestation.

In a similar manner, the age of eggs taken from the uterus may be ascertained by reckoning forward from the time of

[1] E. Van Beneden. Recherches sur l'Embryologie des Mammifères. Archives de Biologie, I., pp. 138-139, 146-153, 1880.

copulation. The fecundation of the egg takes place generally 9 hours after coition, and the first cleavage appears from 10 to 12 hours later.

Method of Taking the Eggs. — After opening the abdomen along the white line, a ligature is fastened around the vessels running to one of the uterine tubes, and then the uterus is cut out. The abdominal cavity is then quickly closed by means of a few stitches. If the operation is performed with care, the rabbit endures it very well, and development progresses normally in the uterus which is left in place. The second uterus may be cut out several hours later, and from it more advanced stages of development may be obtained.

As soon as the first uterus is removed, it is opened by an incision passing from the vaginal extremity along the middle of the side opposite the insertion of the peritoneal fold.

It requires some experience to be able to find the eggs readily during the first few days of their sojourn in the uterus; while from the fifth day onward they become sufficiently large to be easily seen with the naked eye. For a short time after entering the uterus, the eggs are all found near the termination of the oviduct, and are best discovered by searching with a lens between the folds of the mucous membrane. The egg is at this time enveloped by a thick layer of albuminoid substance, which makes it easier to find. From the moment of entering the uterus, the eggs, which are already in the metagastrula stage, begin to increase in size, and at the same time to take up positions at longer and longer intervals, until at length they are ranged along the entire length of the tube.

As the eggs are found, they are raised on the point of a scalpel, and placed on a slide in *aqueous humor* or some other fluid. By the end of seven or eight days, the eggs become so firmly attached to the epithelium of the mucous membrane that they can no longer be detached in an unmutilated condition.

Method of Hardening. — According to Van Beneden, the best method of treatment for the stages of cleavage, the metagastrula, and young blastocysts, is as follows: —

1. Place the egg on a slide in a drop of 1 % osmic acid, and leave it a few moments only.

2. Remove the egg on the point of a scalpel to another slide provided with a few drops of Müller's fluid (or bichromate of ammonia).

3. After an hour renew the fluid, and then leave the preparation 2–3 days in a damp chamber.

4. Transfer to a drop of very dilute glycerine, then to a stronger glycerine, and finally mount in pure glycerine.

For surface views, treatment with nitrate of silver ($\frac{1}{8}$ %) $\frac{1}{2}$–2 minutes gives good preparations.

Method of Preparing Somewhat Later Stages.[1] —

1. After removing the uterus, as before, cut it into as many segments as there are inclosed embryos, taking care to cut transversely midway between two swellings.

2. Place one of the pieces in a dissecting trough, and pin it in such a position that the mesometrical side rests on the bottom of the trough. The convex surface is then turned towards the operator.

3. Whatever be the fluid in which the operation is performed, the egg is opened by seizing the wall of the uterus near the middle of the convex side of the swelling with forceps, and then making a small incision at this point. From the opening a quantity of fluid escapes, which produces a precipitate the moment it comes in contact with picro-sulphuric acid.

4. After removing the coagulated mass from the preparation, one lip of the opening is raised by the aid of forceps, and then a crucial incision of the uterine wall is made with a sharp, fine pair of scissors. By pinning back the four strips, the embryo is brought into view. The placenta is always on the mesometrical side, so that one is certain, in operating as above directed, not to injure the embryo. In order to avoid cutting the terminal sinus of the vascular area, the crucial incision must not be extended too far.

5. By using Kronecker's artificial serum, at a temperature of the body, the embryo may be kept alive for hours.

---

[1] E. Van Beneden and Charles Julin. Recherches sur la formation des annexes fœtales chez les Mammifères. Archives de Biologie, V., p. 378, 1884.

The liquid that escapes when the incision of the wall is made comes from the blastodermic cavity.

THE USE OF COLLODION IN SECTIONING BRITTLE EGGS.

The crumbling of sections of brittle eggs imbedded in paraffine may often be prevented by adding a thin film of paraffine with a hot spatula before making each cut, as recommended by Born (see p. 161); but this can often be accomplished with greater ease and certainty by the use of collodion. The following is a detailed account of the collodion method, as employed by Mark: —

A small bottle, covered with a closely fitting glass cap, is kept about half full of thin collodion, into which dips the tip of a camel's-hair brush. The quill of the brush is thrust through a hole in a thin, flat cork, which rests on the mouth of the bottle, inside the glass cap, and which serves at once as a temporary cover and as a support to the brush, the latter being adjusted to any height of the collodion by simply pushing it up or down through the cork.

The collodion should be quite thin, and used so sparingly that *it will dry almost instantly* (within 2–3 seconds after being applied) *without leaving a trace of glossiness on the surface of the paraffine*. As soon as it begins to leave a shining surface, it should be thinned with *ether*, which is kept in a small bottle near at hand.

Method of Using. — 1. Place the object, imbedded in paraffine, in the receiver of the microtome, and cut away the paraffine to within 1–2^{mm} of the object, leaving a rectangular surface, as in the ribbon method.

2. Lift the brush, wiping it on the mouth of the bottle, in order to remove the most of the collodion, and then paint the paraffine and object by *quickly drawing the brush across the surface*, taking care that the brush be evenly applied, and that the collodion is not carried on to the vertical faces of the block.

3. A moment after the temporary moistening vanishes (within 5 seconds after painting), the knife may be drawn and returned, leaving the section on the edge of the blade.

4. Paint again, and leave the surface to dry while removing the section already cut to a slide prepared with Schällibaum's mixture of collodion and clove oil. Place the section upside down, so that its painted face comes in contact with the fixative.

5. Make the second section, and repeat the steps in No. 4.

If the precautions before named are observed, it will not be necessary to wait for the drying of the collodion film, and the section may therefore be cut at once, — *i. e.*, within 5 seconds after painting. It is thus possible to cut as fast as one can paint the surface; and with some practice it becomes possible to cut continuous ribbons of sections which may be transferred at intervals. It will be found most convenient to cut enough for one row, or half a row, of sections at a time, and then transfer to the slide, rather than to cut the whole object without interruption.

Precautions. — 1. Especial care should be exercised to prevent the painting of the vertical face nearest the operator, since the section is then liable to cling along its whole edge to this vertical film, and be carried under the knife-blade. If by chance this should occur, the section should be removed from the block *before the knife is shoved back*, as it is liable to be caught and lacerated between the face of the block and the under surface of the returning blade. The possibility of the section being thrown under the knife-blade may, however, be obviated, either by carefully trimming (to allow for which the nearer margin of the paraffine may be left broader than the other three) the vertical face, in case it is accidentally painted; or by drawing the knife *slowly*, so that the first indication of a failure to cut through the vertical film may be recognized, and the section held in place on the blade by a slight pressure with a soft brush, whereupon the knife will cut through the film and leave the section free.

2. If, by chance, the paraffine block has been painted with too much collodion, or with collodion that is too concentrated, thus leaving a shining surface, the film should be at once broken by pressing it gently two or three times in quick succession with the *end* of a rather stiff, blunt, *dry* brush. This enables

the collodion to dry quickly, and thus prevents the softening of the paraffine.

3. If the sections have a tendency to curl, they may be flattened out on the slide by means of a brush; for a section thus impregnated with collodion may be handled during the first few seconds after contact with Schällibaum's fixative with much greater freedom than one not so treated. If the sections are cut sufficiently thin (e. g., 5 μ); there is no curling.

This method is useful in preventing the folding of thin sections, as well as the crumbling of brittle ones.

METHODS OF TRACING THE DEVELOPMENT OF THE TELEOSTEAN BRAIN.

I. Rabl-Rückhard's Method.[1] — The eggs are killed in a 10 % solution of *nitric acid*, and the membrane removed as soon as possible (about 15 minutes after immersion), before its contraction has brought it into contact with the embryo. If this precaution is neglected, the embryo is liable to become flattened and otherwise deformed.

After remaining in the acid an hour or more, the preparation is transferred to a 1–2 % solution of *alum* for the same time, and then hardened in the usual grades of alcohol.

The embryo is in the most favorable condition for drawing on the day of preparation. Later, it becomes cloudy white, and the details of configuration are not so easily made out. For drawing, the embryo is placed on a black ground, and illuminated by artificial light concentrated by a condenser.

II. Goronowitsch's Method.[2] — *For surface views*, the egg is immersed in a 5 % solution of *nitric acid* until the outlines of the embryo begin to be visible through the membrane (about 3 minutes), and then removed, before coagulation of the yolk, to a 5 % solution of *alum*. After the lapse of an hour, the yolk appears quite transparent, the embryo alone appearing white.

While in the alum solution, the egg membrane is cut, and carefully removed from the embryo.

---

[1] Arch. f. Anat. u. Entw'gesch., 1882, pp. 67, 68.
[2] Morph. Jahrb., X., H. 3, p. 381, 1884.

The object thus prepared is ready for examination, which was usually made by the aid of artificial light.

In case it becomes desirable to mount the object for later study, this can best be done in 10 % glycerine, to which a little sublimate has been added.

For sectioning, the egg is treated with Kleinenberg's picro-sulphuric acid *three hours,* then passed through 40, 70, and 90 % alcohol.

It is very important to remove the membrane as early as this can be done without injury to the embryo. The operation may be undertaken usually at the end of ten minutes, at which time the embryo begins to be seen through the membrane.

The preparations were imbedded in paraffine, and cut with Thoma's microtome.

CHAPTER IX.

TIMES AND PLACES OF OVULATION.

The Rock-Barnacle (*Balanus balanoides*). — Extremely abundant along our rocky shores, sometimes completely whitening the rocks for long distances. Also very common on the piles of wharves and bridges. Faxon found Balanus larvæ July 10.

Hyperia. — This very interesting little amphipod, which lives in jelly-fishes (*Cyanea*), was found with eggs, at Newport, from May 24th to September. The eggs are transparent (E. N. Whitman).

Phronima. — Only an occasional visitor at Newport; usually found in company with *Salpa*. It is quite transparent, and lives in a barrel-shaped case, open at both ends. The case agrees in size and general appearance with *Salpa*. The transparent eggs are placed in the middle of the case, around the inner surface, forming a more or less perfect ring (E. N. Whitman).

The Prawn (*Palæmonetes vulgaris*). — Abundant in the eel-grass, and especially in the estuaries, occurring far up in the mouths of rivers. Carries eggs in July and August (Faxon).

Hippa emerita. — Lives in shallow water on sandy bottoms. It burrows head first, instead of backward. Eggs produced in July and August (Faxon).

The Lobster (*Homarus americanus*). — "There is a great difference in the breeding season on different parts of the coast. The lobsters from New London and Stonington often lay their eggs as early as the last of April or first of May; while at

Halifax Mr. Smith found females with recently laid eggs in September. At Eastport, Me., the females carry their eggs in mid-summer. In the male, the genital orifices are in the bases of the last pair of legs; in the female, they are at the bases of the middle pair. This will always serve to distinguish the sexes; but they also differ in the structure of their abdominal appendages" (Verrill).

The Green Crab (*Carcinus mœnas*). — Very common along our coast, from Cape Cod to New Jersey (Verrill). The ground color exhibits all shades between a bright green and a dark brown, and is varied by spots and blotches of yellow. The bright orange-colored eggs, carried under the tail, were found at Newport from May 28 to July 20 (E. N. Whitman).

The Rock Crab (*Cancer irroratus*). — Usually larger than the preceding, and easily distinguished by its dull reddish color, sprinkled over with darker, brownish dots, and by having nine blunt teeth along each side of the front edge of its shell (Verrill). One or two specimens were found with eggs in June, at Newport.

The Mud Crab (*Panopeus Sayi*). — I found this crab with eggs throughout the month of June; and Faxon found eggs nearly ready to hatch as late as August 28.

The Blue Crab (*Neptunus hastatus*). — Common on muddy shores, in brackish streams, and estuaries. Easily distinguished from other species by the sharp spine on either side of the back. Eggs found by Faxon August 16.

Occurrence of Eggs at Annisquam. — The following notes on the occurrence of eggs at Annisquam have been kindly furnished by J. S. Kingsley: —

Lineus lays its eggs in masses, surrounded by a glairy matrix, June 1 to June 10. The eggs undergo a regular development without metamorphosis. From three to ten eggs in a mass, eggs slightly whitish, matrix colorless and transparent.

Meckelia ingens contained ripe eggs July 12.

Tergipes despectus lays its eggs in rounded masses of from ten to thirty eggs, attached to hydroids, June 26. The mass is composed of a transparent, colorless jelly, in which the white eggs are closely packed.

Molgula manhattensis. The eggs, in the earliest stages and up to the tadpole stage, were found in the cloaca June 1st; the same stages and later ones about the middle of July of the same year, indicating a breeding season of some extent. The eggs, embryos, and adults live readily in confinement, and seem to require no change of water. Have kept them for six weeks.

Crangon vulgaris oviposits from July 5 to the middle of August.

Gammarus ornatus oviposits the first two weeks in June. The eggs are best obtained by taking couples attached to each other. They live well in confinement, and the young can readily be hatched.

Rossia lays its eggs in water of several fathoms. The eggs are laid singly, usually in the cavities of a large yellow sponge, but sometimes freely on the bottom. The eggs are about the size of a French pea, are spherical, and dark-colored. We found them at Annisquam the first week in August. For other details, see Verrill in report on Cephalopods of the Northeastern Atlantic in Rep. U. S. F. C. for 1878.

Loligo pealii oviposits from at least July 5th to the 15th. (For description of capsule, *vide* Brooks & Verrill.) The capsules are attached below low-water mark, but are frequently thrown up by the tide. They withstand exposure to the air over one tide perfectly, but do not live in confinement long after hatching. During the limits given above they are abundant on Long Beach (Coffin's Beach) opposite Annisquam. I have not had good results with sections. They may be readily freed from the matrix and the egg envelopes with needles. The great difficulty lies in the fact that the thin germinal portion readily separates from the yolk during the process of hardening. The yolk cuts tolerably well, and does not crumble badly. I have had as good results with alcohol of successive grades as by any mode of treatment.

FISH EGGS.

I am indebted to John A. Ryder for the following notes and suggestions:—

The Cod.—The eggs of the cod may be obtained from the

living fishes at Fulton Market wharves during the winter, and artificially fertilized. The ova of the cod are very transparent, without oil drops; the germinal disk is formed very slowly, and occupies an inferior position in relation to the lighter vitellus. They measure about $\frac{1}{20}$ to $\frac{1}{18}$ of an inch in diameter. The microscope must be so arranged as to enable the student to study the living germ disk from below, from the side, and from above. The instrument, when using the camera, should be utilized in the upright and horizontal position, so as to get views of optical sections. In order to get surface views from below, a Nachet inverted microscope or an inverting prism must be used, as the vitellus and germ cannot be kept in the position desired, or be turned, on account of the tendency of these two parts to arrange themselves, the former above and the latter below the other, without such violence as will injure the egg. This is an excellent type in which to study the early phenomena of development, or the changes which take place during fertilization.

The Shad.—*Alosa sapidissima.* Anadromous shad come from the sea to the weedy flats of rivers to spawn. When captured in the seines, the eggs may be taken from the female fishes and artificially fertilized with the milt pressed from the male. Vitellus, $\frac{1}{18}$ inch in diameter; egg membrane, when it has absorbed water and the egg is fertilized, measures about $\frac{1}{4}$ inch in diameter. Vitellus not very transparent. Hatches in three to five days. This is one of the most accessible forms of fish ova for the embryologist. Germ disk lateral in position, egg heavier than water. May be hatched artificially in a jar with a central inlet tube reaching to near the bottom, and an outlet tube reaching a little way through the cork, a gentle current being allowed to pass through the vessel, as shown in the diagram below, the current being regulated by a single stop-cock in the inlet or feed-pipe.

Other heavy ova, such as those of the trout, salmon, and white-fish, are easily hatched during the colder part of the year in this apparatus.

The eggs of *Belone longirostris*, $\frac{1}{4}$ inch in diameter, and those of *Chirostoma* half as large, are provided with filamentary appendages, by which they are fixed to stationary objects in the water. The eggs of the former are found adhering in clusters to the meshes of the pound nets in the southern part of the Chesapeake Bay in July and August. The ova of *Chirostoma* or *Menidia* may be taken artificially about the middle of July. Both are rather troublesome to hatch by means of artificial methods.

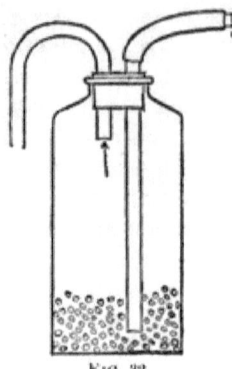

FIG. 32.

The adhesive eggs of the File or Fool-fish, *Monacanthus broccus*, may be taken at the pound nets during the middle of July. These ova are very small and greenish, and contain scattered oil-globules.

Stickleback.—The four-spined stickleback, *Apeltes*, may be obtained, in the month of April, from weedy ditches, and transferred in single pairs, male and female, to aquaria, where, if they are provided with live crustacean food, Copepoda, and water plants, the male will build his nest and place the eggs therein and hatch them out. The eggs are amber colored, a line in diameter, and adhesive, and are easily hatched out artificially in dishes, if the water is simply changed two or three times a day. The eggs will stand a great deal of handling during development, and are to be commended to the student mainly for the purpose of sectioning, as the vitellus is too opaque for the purposes for which the cod's egg is suitable, though much may be observed in relation to the development of the vitelline circulation which develops in a remarkable if not in a unique manner.

Gambusia.—The viviparous minnow, *Gambusia patruelis*, met with in the little streams connecting with the Chesapeake, in the vicinity of Cherrystone, Va., is an interesting form. The female is found with gravid ovaries about the middle of August. It is easily captured with a dip net; is about $1\frac{3}{4}$ inches long, while the male is only a little over an inch. The whole fish may be hardened and the embryos studied in sections as they lie in the ovarian follicles, which latter have a very large opening at one side, which answers to the micropyle of ordinary fish ova.

The Yellow Perch.— *Perca flavescens* lays its eggs in ribbon-shaped masses in April. The egg is remarkable for its complex membrane, which is composed of a thin, outer adhesive layer, a very thick middle canaliculated layer, and an inner zona radiata.

The White Perch has small adhesive ova, which may be pressed in the mature state from the parent fishes in April and May, and hatched on threads which have been drawn through the freshly extruded spawns, the threads being suspended in a hatching jar, through which water is kept flowing.

Gold-Fish.— The eggs of the common gold-fish (*Carassius*) may be had during the months of May, June, and July, in an aquarium provided with aquatic plants, upon which the fishes, if in good condition, will attach their adhesive ova singly. This cyprinoid seems to breed for a considerable period during the warm months.

Cat-Fish.— The cat-fish (*Amiurus*) under favorable conditions will spawn in an aquarium about the middle of July. The ova are $\frac{1}{8}$ inch in diameter, adhesive, in a mass; each egg has an outer adhesive envelope and an inner zona radiata, separated by a thin space from the outer envelope, but connected with the latter by irregularly disposed columns of adhesive matter. The male hovers over the eggs and forces the water through them. They hatch in six days, but the vitellus is not wholly absorbed until about two weeks have elapsed. This form is remarkable for the suddenness with which the ordinal characters of the type appear in the embryo, the barbels, in fact, grow out on the second day of development. It is also remarkable in another respect, viz., the presence of fine, loose granules between the vitellus and the egg-membrane, these granules being stirred about in the breathing chamber of the egg as soon as the embryo is far enough developed to move its tail about. The egg-membranes are exceedingly elastic.

Myriapods.— *Geophilus? Strigamia?* I have found the

nests of these creatures as little burrows or cavities under logs and clumps of loose, damp, friable turf in low, swampy flats, along the mouth of the Susquehanna River, near Havre de Grace, Md. The female was apparently guarding a cluster of about 20 beautiful amethyst-colored ova, each about $\frac{1}{16}$ inch in diameter.

Polyxenes fasciculatus. I have found this neat little myriapod in the chinks of the bark of the sycamore and red cedar in Fairmount Park, Philadelphia. Its eggs are found in the same situation in April and May amongst the detached caudal bristles and ecdyses of the adults as very small oval bodies. It is also found in abundance along the beach of the Chesapeake Bay in St. Mary's County, Md., under fragments of cedar bark; this form is much paler than the inland form, and may be another species.

CHAPTER X.

NUCLEI (KARYOKINETIC FIGURES, &c.).

FLEMMING'S METHOD.

The method employed by Böttcher and Hermann of *overstaining* objects with aniline dyes, and then removing the color to any desired extent by the aid of alcohol, formed the starting-point of some of the methods recently published by Flemming. The following is a summary of the more important of these methods:—

A. For Nuclei in general.[1]—1. Objects hardened in chromic acid ($\frac{1}{10}$–$\frac{1}{2}$ %).

The concentration of the acid and the time of immersion vary according to the character of the tissue. The process of staining is not influenced perceptibly by the time of action of the acid; but objects become very brittle if left a long time in the acid.

2. Carefully washed in distilled water.

3. Transferred directly, in small, thin pieces or sections, to a small quantity (1^{ccm}) of the dye, and left (covered) 12–24 hours.

Safranin, *Magdala red* (rose de naphthaline), and *dahlia* (monophenylrosanilin) give the best staining. Safranin prepared as given on p. 49; magdala in the same way; dahlia best dissolved in water or acetic acid.

Only very small objects or thin sections can be successfully stained.

4. Dye with the object placed in a shallow watch-glass; the watch-glass, with contents, placed in a dish of 70 % alcohol, and agitated a little, in order to remove most of the free dye; then

---

[1] Flemming. Archiv f. mik. Anat., Vol. XIX., p. 321, 1881.

transferred at once to absolute alcohol for half a minute or longer, till no visible clouds of color are seen about the object.

The process of decoloring is now completed, and the objects must be at once removed from the alcohol; otherwise, the color will be too much weakened. If it be required to examine the objects before mounting, they may be removed to distilled water, in which the color of the nuclei will remain unchanged for a considerable time. They must then pass through alcohol again before mounting.

As the aim in decoloring is to bring out a strong, differential staining of the nuclei, the process should be checked the moment it reaches the nuclei. No precise time-limits can be given, as the length of the process varies with the thickness and kind of tissue.

5. Clarified in clove oil, and mounted in *damar-lac*.[1]

Clove oil withdraws the color a little, and hence it must not be allowed to work too long. Creosote extracts the color still more rapidly.

B. Eggs of Echinoderms.[2] — In somewhat later researches on karyokinesis, Flemming states (p. 5) that he obtained serviceable staining of nuclei in the two following ways: —

1st. *Living eggs colored on the slide,* either with *safranin* or *aniline dyes,* followed by acetic acid (1 %), which is allowed to flow under the cover and thus replace the staining medium, or with *Acetic acid carmine* (after Schneider) used undiluted.

The last mentioned staining agent causes swelling, but still gives the typical features of the karyokinetic figures.

2d. Eggs first hardened in strong nitric acid (40–50 to aq. dest. 60–50), then washed in distilled water until the yellowish color, due to the presence of the acid, disappears. Colored with acetic acid carmine.

C. Chrom-Osmium-Acetic Acid and Safranin.[3] — Among the various methods of studying karyokinetic figures ("mitosis") which have been given by Flemming, the latest, most detailed, and most highly recommended is the following: —

---

[1] Flemming prepares this by dissolving (warm) one part of gum damar in one part of turpentine and one of benzole, and then evaporating to a syrupy consistency.

[2] Arch. f. mik. Anat., XX., p. 1, 1881.

[3] Zeitschr. f. wiss. Mikroskopie, I., Heft 3, pp. 349–361, 1884.

1. Place fresh-cut pieces of tissue (best not more than $\frac{1}{2}^{cm}$ thick) in a mixture of

    Chromic acid (1 %) . . . . 15 parts
    Osmic acid (2 %) . . . . 4 "
    Glacial acetic acid . . . . 1 part (or less)[1]

They should remain in this fluid *at least* one day; for complete hardening, better 2–3 days, or even weeks or months. They are not injured by exposure to light.

2. Wash in water for an hour or more. This is best done by placing the object, enclosed in a small box of wire netting, in a strong current of water.

3. Harden in absolute alcohol a few hours.

4. Imbed, *while still saturated with alcohol*, in very soft paraffine, elder pith, or in celloidin.

This method of imbedding is considered best and safest for obtaining clear and well-preserved nuclear figures; but, as Flemming remarks, it is not to be adopted in cases where it is important to have a uniform series of thin sections.

5. Cut with the knife flooded with alcohol.

6. Wash the sections in water, and then transfer them to a strong solution of safranin (or gentian violet) for 24 hours or more.

7. After washing in water, place in absolute alcohol to which a little hydrochloric acid ($\frac{1}{2}$% or less) has been added, leaving them for *only a few moments*, until clouds of color are no longer seen.

8. Transfer to pure absolute alcohol, and finally clarify with clove oil, and mount in damar or balsam.

Special stress is laid on this method as a convenient means of finding karyokinetic figures, and of determining, by their frequency, *places of most rapid growth*.

STRASBURGER'S METHODS.[2]

In the study of nuclear metamorphosis, Strasburger made use of 1 % *acetic acid*, with which a little *methyl green* was

---

[1] A much stronger mixture than was recommended in his "Zellsubstance, Kern, and Zelltheilung, 1882, p. 381.

[2] Arch. f. mik. Anat., XXI., p. 476, and "Zellbildung und Zelltheilung," 3d ed., p. 141.

mixed. The figures are "fixed" and colored at the same time, but the preparations are not permanent.

Preparations were also made by treating with alcohol, or 50 % nitric acid, and then staining with the coloring substance mixed with dilute glycerine. Rapidity of preparation and clearly defined spindles compensate, in these methods, for the lack of permanency.

RABL'S METHODS.[1]

Material. — The skin and kidney of Proteus and the epithelium of the mouth of Salamander larvæ. The epithelium is the more favorable object, as the very large nuclei can be examined in surface preparations. The achromatic spindles are seen to best advantage in the renal tissue.

Preparation. — *a.* Place small, fresh pieces of the object in *chrom-formic acid* (200 g. of a $\frac{1}{3}$ % solution of chromic + 4 to 5 drops of strong formic acid) 12–24 hours.

b. Wash thoroughly, and harden slowly, first 24–36 hours in 60–70 % alcohol, then in absolute alcohol.

Instead of the chrom-formic acid, a $\frac{1}{3}$ % solution of *platinum chloride* may be used; the preparation otherwise remaining the same. Chrom-formic acid causes the chromatin filaments to swell somewhat, so that their longitudinal division generally becomes obliterated; while platinum chloride causes a slight shrinkage which brings out very distinctly the division of the filaments, as well as the chromatin spherules of Pfitzner.

c. Stain in either of the three following ways: —

1. *Grenacher's hæmatoxylin* (strongly diluted with distilled water) 24 hours, followed, after washing, with acidulated alcohol (few drops of HCl.).

2. *Pfitzner's safranin* 2–4 hours, followed by absolute alcohol, in which the object is left until no visible cloud of color remains upon turning it over (generally about two minutes), clove oil a few minutes, and damar.

3. Double stain with hæmatoxylin and safranin; stain very feebly with hæmatoxylin; wash and treat with acidulated alcohol; and then stain with safranin as in No. 2.

---

[1] Morph. Jahrb., X., H. 2, pp. 215–219, 1884.

NUCLEI (KARYOKINETIC FIGURES, ETC.).   185

Examination. — High powers are required in the study of the mounted preparation, either the homogeneous immersion $\frac{1}{18}$ of Zeiss, with Abbe's condenser, or that of Hartnack, No. III., $\frac{1}{24}$. Nachet's camera was employed in drawing.

It is well to work with green light, which can be obtained by inserting a green-colored glass plate beneath the table of the microscope, as was first recommended by Engelmann.

The slide devised by Rabl enables one to examine a preparation from both sides. It consists of four pieces of glass, of the shape and size seen in the figure (*a, b, c, d*), and a cover-glass, *g*, which serves as the object-bearer. The two glasses, *a* and *b*,

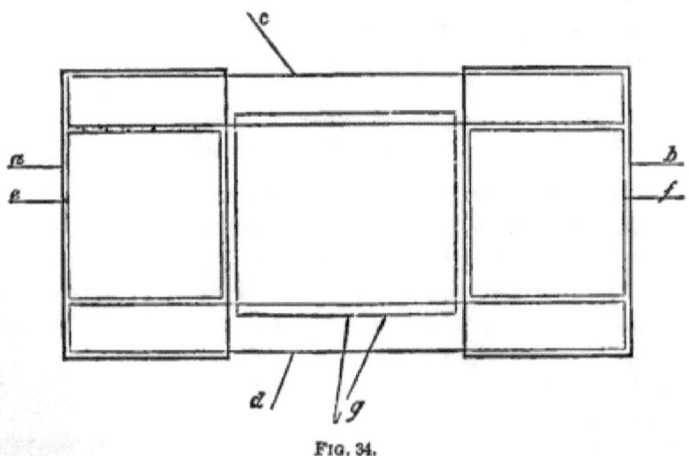

FIG. 34.

are painted on one side with chloroform balsam, and then connected by means of *c* and *d*. The frame thus formed is completed by adding the pieces, *e* and *f*, between *c* and *d*. The frame is next turned over, and the middle portions of the glass bars, *c* and *d*, painted with balsam; and a thin glass cover, *g*, placed so as to rest on the painted sides of *c* and *d*. This glass (*g*) bears the object in damar, which is covered by another very thin glass. The object, lying between two thin cover-glasses, can be viewed from both sides with the highest powers.

HEIDENHAIN'S METHOD OF USING HÆMATOXYLIN.[1]

1. Pieces of organs are first well hardened in alcohol.
2. Transferred to a ½–1% aqueous solution of hæmatoxylin 8–10 hours.
3. Treated with bichromate of potassium (½–1%) 8–10 hours.
4. After becoming thoroughly black, they are washed in water, dehydrated with alcohol, imbedded in paraffine, and mounted in balsam.

Nuclei are thus stained black, while other parts of the cells appear dark gray, or likewise black. Sections require to be very thin.

This method of staining may be followed equally well with sections (Minot).

DEMONSTRATION OF NUCLEI IN THE PSEUDO-CHLOROPHYLL BODIES (ZOOCHLORELLA) OF HYDRA.[2]

1. Color the green corpuscles alive, after setting them free by pressure, in an aqueous solution of hæmatoxylin. Remove the excess of coloring substance by washing in water, and then kill in absolute alcohol. The chlorophyll is generally extracted in about two hours, at the end of which time the nuclei, one or more in each cell, appear as violet-colored dots.

2. First kill with osmic or chromic acid (⅕%), then stain with hæmatoxylin.

3. (After Flemming). Leave pieces of Hydra in ⅕% chromic acid 1 hour; wash, and stain with a solution of magdala in 50% alcohol (24 hours). Place the pieces in absolute alcohol, several times renewed, in order to remove a part of the color. Isolate the green cells (in water or balsam) by pressure. Nuclei are colored as before.

---

[1] Arch. f. mik. Anat. XXIV., H. 3, pp. 468–470, 1884.
[2] K. Brandt. Arch. f. Anat. u. Phys. phys. Abthg., 1882, pp. 133–134.

CHAPTER XI.

PREPARATION OF NERVOUS TISSUE.

GOLD CHLORIDE METHODS.

Flechsig's Method. — 1. Place the organ (brain, oblongata, spinal cord, etc.,) in a 1 % solution of ammonium bichromicum, and let it remain until it becomes hard enough to section (about 6 days).

2. Wash the sections for a short time in distilled water, and then place in a $\frac{1}{2}$% solution of gold chloride 15–30 minutes.

3. Transfer, after washing again in distilled water, to a solution of caustic soda (10% or less), in which the white matter immediately assumes a dark violet color, while the gray matter remains apparently uncolored. A few hours to one or more days.

4. After washing, treat with alcohol, and mount in the usual way.

One of the most reliable methods of using gold chloride is that employed by Dr. Sigm. Freud,[1] in investigating the course of the nerve-fibres in the oblongata of the human fœtus.

Freud's Method. — 1. Harden in Erlicki's fluid ($2\frac{1}{2}$ parts of bichromate of potash, $\frac{1}{2}$ sulphate of copper, and 100 aq. dest.) until the preparation has become sufficiently firm for cutting; then transfer to alcohol, and section at convenience, taking care only that this be done *before* it becomes brittle.

2. Wash the sections well in distilled water, in order to free

---

[1] Arch. f. Anat. u. Physiol., Anat. Abth., 1884, pp. 453–460.

them from the alcohol with which the knife of the microtome is wet, and place them in a 1% solution of gold chloride 3–5 hours. If the sections are thoroughly washed, no film or precipitate will form in the gold chloride, and the same solution will bear repeated use.

3. After washing, place in strong caustic soda (1 part *natrum causticum fusum*, and 5–6 parts water) 3 minutes. The sections, for the most part, become at once transparent.

4. Remove the sections, one at a time, to a glass slide, and free them from the lye, so far as this can be done by a careful use of blotting-paper.

5. Place them in a 10–12% solution of iodide of potassium 5–15 minutes, in which they flatten out, and gradually assume a red color in consequence of a reduction of the gold.

It is well to have the sections as free as possible from the lye before placing them in the iodide solution, and to renew the latter after staining from 10 to 25 sections.

6. Place the sections again on a glass slide, and dry them by the aid of filtering paper; loosen them from the slide by dipping it in a glass of distilled water; change the water; treat with the ordinary grades of alcohol; then clarify and mount in the usual way.

This method can be varied to advantage, in many cases, by adding to the 1% gold-chloride solution an equal volume of 94% alcohol.

Gaule's Method for Nerves. — 1. Object treated with chromic acid ($\frac{1}{4}$%) 12 hours.

2. Washed 12 hours.

3. In 1% gold chloride (dark) 1 hour.

4. Washed in distilled water, and placed in formic acid 24 hours (or until a violet color is obtained).

5. Washed and treated with alcohol.

6. Imbedded in paraffine.

Ranvier's Methods for Nerve-Terminations. — *A.* 1. Muscles of salamander placed direct in a mixture consisting of: —

> Gold chloride (1%) . . . . 4 parts
> Formic acid . . . . 1 "

The two fluids are mixed, heated to boiling point, and used immediately after cooling. This process renders the gold chloride more easily reducible.

2. Washed at the end of 20 minutes in

Formic acid . . . . 1 part
Water . . . . 4 "

B. 1. Muscle of frog placed in fresh filtered lemon juice 5-10 minutes.

2. Washed in water, and then left 20 minutes in 1% gold chloride.

3. After washing again, placed in acidified water (water 50 cc. + acetic acid 2 drops) in daylight for 24-48 hours.

Such preparations grow darker with time.

In the method of Loewit,[1] the object was first treated with formic acid before being exposed to the action of gold chloride; the result was that the terminal ramifications of the nerves were deleteriously affected by the first reagent before they could be fixed by the second. In combining the two, as in *A*, Ranvier hoped that the gold chloride would accomplish its work before the nerves were injured by the acid. The result, although not quite satisfactory, showed that the method was far better than that of Loewit. As the injurious effect of formic acid could not be wholly avoided, Ranvier attempted to replace it by a more favorable acid, which he found in lemon juice. This led to method *B*.

Motor Nerve Endings. — Ciaccio[2] has investigated the motor nerve-plates in the depressor muscle of the jaws of *Torpedo marmorata* by treatment with double *chloride of gold and cadmium*. From the anterior third of the muscles, strips one millimeter thick were cut with scissors, the strips were then left for five minutes in fresh filtered lemon juice, washed in distilled water, and placed for half an hour in 1% solution of gold and cadmium, being kept dark; washed again in 1% aqueous solution of formic acid, in which they were left 12 hours in the dark, then 12 in the light; finally, kept in the

---

[1] Cf. Quart. Jour. Micr. Sc., XX., No. 80, pp. 456-458, 1880.
[2] Arch. Ital. Biol., III., 75.

dark in stronger formic acid for one day, and preserved in glycerine. The fibres of such strips may easily be dissociated.

Pfitzner's Method of Studying Nerve-Terminations.[1] The material for investigation consists of a fresh piece of the epidermis (circa 6^{mm} in diam.) taken from the ventral (unpigmented) side of a well-fed salamander larva. The methods of treatment are as follows:—

1. Hardened in chromic acid ($\frac{1}{4}$–$\frac{1}{3}$ %) for a number of days.
2. Imbedded in elder pith, and cut with the microtome into sections 10–15 μ. thick.
3. Sections washed in distilled water for at least 30 minutes, (not injured by remaining in water 24 hours).

The further treatment of these sections may take either of the following courses: —

A. 1st. Colored with safranin and mounted in damar.
   2d. Colored with hæmatoxylin and mounted in damar.
   3d. Mounted unstained in glycerine.

B. 4th. (a) Transferred to a 1 % solution of *gold chloride* (15–30 minutes) to which has been added a trace of hydrochloric acid. Kept dark.

(b) Carefully washed, half an hour or more, and then placed in 5 % formic acid 12–24 hours (exposed to light).

(c) Again thoroughly washed in distilled water, and then mounted directly in glycerine, or, after staining with safranin, in damar.

C. 5th. (a) Transferred to $\frac{1}{2}$ % osmic acid for 30 minutes or longer.

(b) Carefully washed and mounted in glycerine.

GOLGI'S SILVER METHOD.[2]

For the investigation of nerve-fibres, Golgi has employed with success the following method:—

1. A fresh nerve (rabbit) is placed for one hour in a mixture of

Bichromate of potassium (2 %)	10 parts
Osmic acid (1 %)	2 "

---

[1] Morph. Jahrb., VII., pp. 291, 292, and 731, 1881.
[2] Arch. per le sc. med., IV., p. 221, 1880.

2. The nerve is then cut into pieces ½–1° in length, and again returned to the mixture for some hours.

3. Transferred to nitrate of silver (½ %) for 8 hours.

4. Sectioned and mounted in damar.

Golgi made preparations in still another way: —

1. Peripheral nerves treated with bichromate of potassium 4–8 hours (Central organ 10–15 days).

2. Transferred to silver nitrate 12–24 hours (in dark).

3. Sectioned and mounted as before, and then exposed to light for reduction.

WEIGERT'S METHOD OF STAINING THE CENTRAL NERVOUS SYSTEM WITH HÆMATOXYLIN.[1]

Weigert's hæmatoxylin method is much more expeditious, simple, and easy than his fuchsin method; and, as it gives equally good or even better results, and is far less capricious, it has superseded the latter method.

1. Object hardened in Müller's or Erlicki's fluid.

The hardening process in Müller's fluid may be completed in 8–10 days, instead of 8 weeks, by heating in an incubator kept at $30°$–$40°$ C.

In Erlicki's fluid, the hardening requires ordinarily 8–10 days; by heating it may be accomplished in about 4 days (Centralbl. f. d. med. Wiss., XX., No. 46, p. 819, 1882).

2. Transferred directly, *without ever coming in contact with water*, to alcohol.

If the object is allowed to remain in alcohol until it becomes green, it must be returned to the original hardening fluid, and left until the brown color is restored.

3. Best imbedded in celloidin.

4. The *brown* (not green) sections are stained in

    Hæmatoxylin . . . . 1 part
    Alcohol . . . . .10 "
    Water . . . . .90 "

This staining mixture is boiled and allowed to stand some days before using. The sections remain 24 hours (1–2 hours, if kept at a temperature of $35°$–$40°$ C.) in the dye, but may remain without injury any length of time before being transferred to the decoloring fluid. The sections must not be exposed to water before staining.

---

[1] Fortschritte der Medicin, II., No. 6, p. 190, 1884. Cf. Edinger's Review in Zeitschr. f. wiss. Mikroskopie, I., H. 2, pp. 290–292, 1884.

5. The black-colored sections, after being rinsed in water, are decolored in

 Borax . . . . 2 parts
 Ferricyanide of Potassium (red prussiate of potash) 2½ "
 Water . . . . .100 "

The first traces of a distinction between white and gray substance will appear in about 30 minutes; but the decoloration must be allowed to go on until *the gray substance becomes plainly yellowish*, the white substance remaining black.

6. The sections are next washed thoroughly in water, then treated with alcohol, clarified in xylol, and mounted in balsam.

Max Flesch[1] has made some important modifications of Weigert's method. In the first place, he discovered that preparations which had been washed in water in the usual way, after coming from Müller's fluid, could be stained, provided the sections (made in celloidin) were treated a few minutes in ½ % chromic acid, and then, after being washed in water, placed in the coloring fluid. The sections stain very much quicker than in Weigert's method. The decoloring process of Weigert is followed. Creosote is said to be superior to xylol as a clarifying medium.

According to Minot, Weigert's hæmatoxylin method may be used after any method of hardening and cutting, provided the sections are treated 5–15 minutes in 1 % bichromate of potassium, then washed in water, and transferred to the staining mixture. Instead of bichromate of potassium, the following mixture may be used with equal success, but with somewhat *different* results : —

 Water . . . . 100 cc.
 Alum . . . . 1 g.
 Bichromate of potassium . . . . 1 g.

THE NUMERICAL RELATION OF NERVE FIBRES TO GANGLION CELLS IN THE SPINAL CORD OF THE FROG.

In the very difficult task of determining the precise numerical relations of nerve fibres to ganglionic cells, Prof. E. A. Birge,[2] working under the direction of Dr. Gaule of Leipzig, employed the following methods with success : —

---

[1] Zeitschr. f. wiss. Mikroskopie, I., Heft 4, pp. 564–566, 1884.
[2] Arch. f. Anat. u. Phys., Phys. Abthg., 1882, pp. 437–440 and 446–448.

A. Method of Counting Nerve Fibres. — *a. Preparation.* 1. Lay bare the spinal cord in its whole extent by the aid of bone-nippers; then, after cutting off the roots close up with sharp shears, remove the cord, in order to give the hardening agent free access to the vertebral canal.

2. Cut off the nerves at some distance outside the intervertebral canals, through which they pass, and remove the entire vertebral column, together with the roots and nerve portions, to a 1 % solution of osmic acid, in which, protected from the light, it is allowed to remain six hours. Before treating with the acid, it is well to remove the muscular tissue as far as possible. The pleuroperitoneal membrane covering the nerves should not be disturbed.

3. After washing 2–3 hours in distilled water, place the preparation in 70 % alcohol for 24 hours, then in 96 % alcohol.

4. After dehydration, the nerves with their roots may easily be removed from the vertebral column. In order to avoid confusion, each should be isolated in a small open box, made of tin-foil and provided with a number.

5. Preparatory to imbedding, transfer the preparations from alcohol (each in its tin-foil box) to clove oil, then to turpentine, then to a mixture of turpentine and paraffine, heated to 45° C., for one hour, and finally to pure paraffine, at 55°–60° C., for 20 minutes.

6. Sections of the imbedded nerves should be made as thin as $\frac{1}{200}^{mm}$. They are mounted as they are taken from the knife (without removing the paraffine) in zylol balsam, after Gaule's method.[1]

b. Counting. An examination of the thin transverse sections of the nerves with the microscope shows how important a part osmic acid plays in the above method of preparation. The medullary sheaths are stained black, and appear in sections, as small black rings, at the centre of each of which is seen a minute gray dot (the axis cylinder). Each individual fibre, however small, is thus rendered perfectly distinct. The exact

---

[1] The methods since introduced of fixing sections on the slide would doubtless be more convenient and satisfactory in every way.

number of fibres in the nerve may be determined by the use of a net-micrometer, which divides the section into conveniently small areas for counting.

B. Method of Counting Ganglion Cells. — *a. Preparation.* 1. Having laid bare the central organs of the nervous system, cut off the roots as far as possible from the spinal cord, and then remove the brain and cord to Müller's fluid (7 days).

2. Wash carefully and thoroughly, and then transfer to 70 % alcohol for 24 hours.

3. Stain with Grenacher's alum-carmine.

4. Wash, treat with alcohol, clarify, and imbed in paraffine. Sections need not be thinner than $\frac{1}{75}^{mm}$, and should not be thicker than $\frac{1}{50}^{mm}$. They must be mounted in serial order, and without any loss.

b. Examination. The nerve fibres are colored very little, the gray substance deeply so. The counting of the ganglion cells is comparatively easy, as the number of cells on either side of a section is only from two to ten. The number of cells is tabulated separately for each side and for each section, and the difference between the two sides noted. As the total number of cells varies with the weight of the frog, a record of the weight must also be kept.

BETZ'S METHOD OF HARDENING THE BRAIN.[1]

1. Harden in 70 % alcohol containing enough iodine to give it a yellowish-brown color. After a few days the color disappears, and a little iodine tincture is again added (6–10 days).

2. Transfer to a 3 % solution of bichromate of potassium, and leave until completely hard. The time for removal from the fluid will be marked by the formation of a reddish-brown precipitate on the surface of the preparation.

During the entire process of hardening, the preparation should be kept in a cool place.

THE BRAINS OF URODELA.

The following method of preparation is extracted from Dr.

---

[1] From Fol's Lehrbuch, p. 105.

H. F. Osborn's papers[1] on the brains of American Urodela, and from a letter in which the details were more fully given: —

"Before hardening, the brains were inflated with Müller's fluid, so as to preserve the natural proportion of the cavities. After treatment with alcohol, they were placed for a week in dilute carmine. Calberla's egg-mass was employed as before, except that *the ventricles were injected with the mass before hardening*. The delicate parts of the brain-roof were thus retained. It appears now that celloidin may be used for this purpose to equal, if not to greater, advantage in results, and with considerable economy of time. The sections were cut in absolute alcohol, then floated upon a slide in consecutive order, from

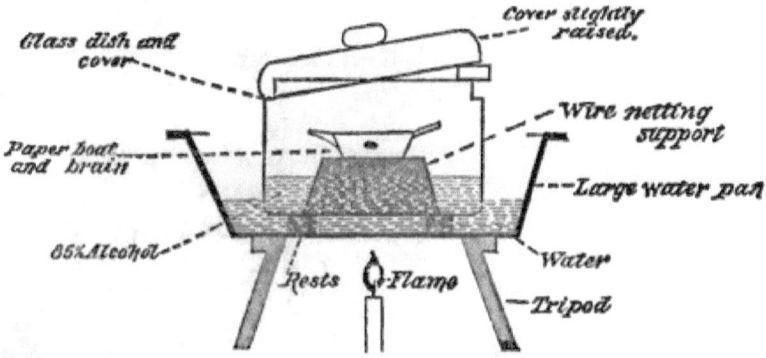

FIG. 35. — Bath for hardening egg-emulsion.

20 to 50 at a time, and covered with a delicate slip of blotting-paper during treatment with oil of cloves."

Imbedding. — 1. The egg-mass was prepared by shaking the white and yolk of egg together, with three drops of glycerine to each egg, and then well filtered through coarse cloth.

2. The bath is then prepared as follows: —

a. Outside is a large water-pan, for boiling with a Bunsen burner, etc.

b. Inside this, supported on rests to prevent jarring, is a covered glass dish, filled to about 1 inch in depth with 85% alcohol.

---

[1] Proc. Acad. Nat. Sc. of Philadelphia, 1883, p. 178, and 1884, p. 262.

c. Within the glass dish is placed a piece of coarse wire netting which supports the imbedding box, raising it above the alcohol.

3. The box, made of paper in the usual way, and one-fourth filled with the imbedding mass, is kept in the bath until the mass is hardened enough to support the brain. The brain is next placed on the hardened stratum, and covered with the fresh mass. The second stratum is hardened just enough to hold the brain in place, and then a third is added, filling the box.

4. The whole mass must now be allowed to harden through and through, requiring about fifteen minutes.

5. The hardening is completed by passing the box through three grades of alcohol, — 80, 90, and 100%, — allowing it to remain 24 hours in each.

When the mass becomes nearly white, and ceases to discolor the alcohol, it is ready for cutting.

MOUNTING AND PHOTOGRAPHING SECTIONS OF CENTRAL NERVOUS SYSTEM OF REPTILES AND BATRACHIANS.[1]

Both the brain and spinal cord were entirely separated from the body, and, with their membranes, placed in iodine-tinted alcohol until they had acquired a slight degree of consistency, — from 6 to 12 hours. They were then transferred to a 3% solution of bichromate of potash, with a small piece of camphor, in a tightly-corked, wide-mouthed bottle, and allowed to remain until ready for cutting, renewing the solution every two weeks.

The time required for the hardening process varies considerably in different animals, and this variation is more dependent upon the class of animal, than upon the relative dimensions of the specimens.

For example, on the same day I placed the brain of a large rattlesnake with that of a small salamander in the same bottle, and at the end of six weeks the former was ready for sectioning, whilst the latter was not sufficiently hard until a month after-

---

[1] J. J. Mason. Minute Structure of the Central Nervous System of Certain Reptiles and Batrachians of America, 1879, 1882.

wards. By thus employing the same reagent in all cases, I have been able to note constant differences in the action of both the hardening and the coloring agent, carmine.

Perhaps the most striking illustration of this is furnished by the nervous centres of tailed batrachians, which, while they stain very readily, invariably require about a third more time to harden than specimens from the other orders. Specimens from ophidians stain less satisfactorily than those from any other of the classes which I have studied, while with the spinal cords of alligators, turtles, and frogs, failure to obtain good results in this particular is very rare.

In all cases the sections have been stained after cutting, injury from excessive handling being wholly avoided by the use of siphon tubes to remove the alcohol and washings. For producing transparency, oil of cloves has been used, and the mounting has been done under thin, clear covers, in a solution of Canada balsam in chloroform.

All the negatives have been made on glass thoroughly cleaned and lightly coated with a solution of wax and benzole, so that the collodion film, previously made adherent to thin sheets of gelatine, could be safely removed from the plate. The flexible negatives thus obtained are well adapted to the artotype process, and, as they can be indefinitely preserved between the leaves of an ordinary scrap-book, are very desirable for a series of illustrations. In making the original negatives on glass, the "wet collodion process," with the sulphate of iron developer, has been exclusively employed.

The prints correspond exactly with the negatives, both in outline and detail. No distinction occurs, as in silver printing, in which process the paper is subjected to prolonged washing.

In many of the photographs the gray substance appears lighter in shade than the white substance. This appearance is due to a greater degree of transparency of the gray substance in these sections, resulting from the action of the oil of cloves, followed by an increased action of the transmitted light on the sensitive collodion film of the negative, and hence by a thinner deposit of ink over corresponding parts of the positive plates from which the artotypes are printed.

With regard to the process employed, Dr. Mason says that after experimenting with various methods he found that satisfactory prints could be made in ink directly upon plate-paper, and that these impressions were as perfect in fine detail as any of those obtained by the silver process of printing. The plates (all printed by the autotype process), are as durable as steel engravings. While a photograph often cannot show all that can be discovered by more direct microscopic observation with a judicious working of the fine adjustment, high authority has stated, and perhaps correctly, that a good photograph with a low power, say from three to one-half inch, is a better means of illustrating the anatomical structure of the nervous tissues than hand-drawing. Some of the plates with high powers leave much to be desired both in distinctness and tone, and in general it may be affirmed that the same defect as regards distinctness always exists, and for obvious reasons, in photographs of sections with powers much above one-half inch.

METHODS OF STUDYING THE NERVOUS SYSTEM OF ANNELIDS.[1]

Maceration is the best means of demonstrating the existence of a peripheral nervous system (Polygordius, Protodrilus, and Saccocirrus) and of showing its relation with the central nervous system. As macerating agents, Fraipont employed weak alcohol (36–48 hours), chromic acid ($\frac{1}{100}$ %, 24 hours), and a weak solution of bichromate of potash (48 hours).

After treatment with one of these agents, a definite portion of the annelid may be placed on a slide, teased apart under the dissecting microscope with fine needles, and then examined in a drop of the macerating fluid; or it may be freed from its cuticula, and subjected to gradual pressure under a cover-glass. This treatment causes the preparation to flatten out, but does not dissociate the tissues so far as to obscure the relations existing between the different layers and their constituent elements.

In some cases, good results may be reached by giving light taps on the cover-glass with the point of a needle for ten min-

---

[1] Julien Fraipont. Arch. de Biol., V., F. 2, pp. 251-254, 1884.

utes or more. In either case, the progress of dissociation can be followed with the microscope.

Specimens to be sectioned with the microtome should be so killed that they remain straight and extended. They may be killed by adding very slowly alcohol to the water. As soon as they cease to move, they should be taken out, and extended on a slide, and then hardened with alcohol, osmic acid, picro-sulphuric acid, chromic acid, or corrosive sublimate.

Another method of killing is to pour hot corrosive sublimate over the worm after it has been stretched out on a dry slide. A mixture of osmic acid (1 %) and chromic acid ($\frac{1}{2}$ %) in equal parts was also employed with some success.

For coloring, borax carmine was employed after sublimate and chromic acid; picrocarminate of ammonia after alcohol, osmic, and picric acid; hæmatoxylin and aniline dyes after chromic acid.

CHAPTER XII.

HISTOLOGICAL METHODS.

METHODS OF INVESTIGATING ANIMAL CELLS.[1]

As most Protozoa move very rapidly when hungry, it is well to feed them before attempting to study them with the microscope. If well fed with powdered pieces of plants, etc., they usually remain quiet after a short time, and begin to assimilate the food material which they have appropriated. In this condition of comparative quiet they can be easily examined with high powers. For this purpose they may be placed under a cover-glass with considerable water and a number of small green algæ to keep the water supplied with oxygen.

For higher powers, Abbe's illuminating apparatus is extremely useful. In some cases, it is desirable to have a completely one-sided illumination, and this is best effected by inserting beneath the illuminating apparatus a circular diaphragm-plate perforated with a slit 3^{mm} wide that runs parallel to the edge of the plate. It is best to have about 2^{mm} between the slit and the edge of the plate. Several diaphragm plates should be prepared in which the slit varies in extent from a half to a whole of a quadrant or more.

The following mixture, which is Merkel's fluid with the addition of a little acetic acid, is recommended above all other reagents as a preservative medium:—

Chromic acid	1 part
Platinum-chloride	1 "
Acetic acid	1 "
Water	400–1000 "

---

[1] A. Brass. Zeitschr. für wiss. Mikroskopie, No. I., pp. 39–51, 1884.

Unicellular animals die very slowly in this mixture, and suffer very much less alteration in structure than when killed in osmic acid or picro-sulphuric acid.

A special method is required for Protozoa filled with opaque food material. In many cases the nucleus and the structure of the cell body are completely obscured by foreign bodies. The method adopted in such cases is as follows:—

(1) Placed in picro-sulphuric acid 3–4 minutes.
(2) Transferred to boiling hot water for a short time.
(3) Placed in water, and a little ammonia added; this causes the contracted object to swell up to its original size and form.
(4) Neutralize the ammonia with a little acetic acid, and then
(5) Color with borax-carmine or ammonia-carmine.
(6) Wash and examine in dilute glycerine.

The picro-sulphuric acid destroys the nutritive material; the ammonia dissolves any particles of fat that may be present; and thus the object becomes transparent so far as possible.

A concentrated solution of corrosive sublimate may also be used with success for killing Protozoa; but care must be taken to wash thoroughly.

Dr. Brass has obtained his best results without reagents or dyes.

BLANC'S METHOD OF PRESERVING AND STAINING PROTOZOA.

For preserving these small organisms, various reagents have been recommended. Certes[1] and Landsberg[2] employ osmic acid; Korschelt,[3] chromic acid or osmic acid; and Entz,[4] Kleinenberg's picro-sulphuric acid.

Finally, Blanc[5] recommends the following very dilute picro-sulphuric solution:—

Picric acid (saturated solution in dist. water)	100 vol.
Sulphuric acid (concentrated)	2 "
Distilled water	600 "

To this solution, which may be employed as it is for the larvæ of Echinoderms, Medusæ, and Sponges, a little acetic acid

---

[1] Compt. Rend. Acad. Sc., Paris, t. 88.
[2] Zool. Anzeiger, No. 114.
[3] Zool. Anzeiger, No. 109.
[4] Zool. Anzeiger, No. 96.
[5] Zool. Anzeiger, No. 129.

(1 %) is added for Rhizopods and Infusoria—two or three drops for 15 grams of the solution. The acetic acid is added in order to sharpen the outlines of the nuclei and nucleoli.

This liquid is preferable to osmic acid, because it does not render the objects non-receptive to staining fluids.

The entire process of hardening, washing, staining, and mounting can be more expeditiously performed under the cover-glass than otherwise. The acid is allowed to work until the objects have become thoroughly yellow. The acid is then replaced by 80 % alcohol, frequently renewed until the yellow color entirely disappears; 96 % alcohol is next used, and then absolute alcohol.

The hardened objects may be stained with picro-carmine, or, better, with an alcoholic solution of safranin. Five grams of safranin are dissolved in 15 grams of absolute alcohol; the solution left standing a few days, then filtered, and diluted with half its volume of distilled water.

This solution of safranin is preferable to picro-carmine, because it colors more quickly, and because one can so regulate its action as to give a sharp definition to the protoplasm or the nucleus.

After the object has been more or less deeply stained, according to the end in view, it is washed in 80 % alcohol, which is renewed until a moment arrives when no visible clouds of color appear; at this moment the 80 % alcohol is replaced with absolute alcohol, and this by clove oil.

As safranin is soluble in alcohol, the process of washing will of course remove or weaken the color; but decoloration is gradual, so that one needs only to watch and apply the clove oil when the color has been reduced to the desired intensity. This process, then, consists in *overstaining* and then *removing* the color to any desired degree. The process of decoloration is not entirely arrested by the application of clove oil, contrary to Blanc's assertion, hence it should be replaced by Canada balsam as early as possible. The same method is adapted to other microscopic animals.

METHODS OF PREPARING THE ALCYONARIA

Wilson[1] has employed with success the following methods in preparing the Alcyonaria for histological purposes: —

"After testing many preservative and staining fluids, the following methods were adopted as giving, upon the whole, the best results: The animals were suddenly killed by momentary immersion in a mixture of one part strong acetic acid and two parts of a concentrated solution of corrosive sublimate in fresh water. After being quickly washed, they were transferred to a concentrated solution of sublimate in fresh water, and left two or three hours, the internal cavities being injected with the solution where this was possible. They were then thoroughly washed in running sea-water, then in distilled water, and finally preserved in successive grades of alcohol.

"A weak solution of iodine in alcohol and sea-water also gives beautiful results, but is less certain in its action. For staining, I have used Grenacher's alum-carmine, borax-carmine, picro-carmine, and Kleinenberg's hæmatoxylin. Much the best results are obtained by the use of alum-carmine; but it must be used as quickly as possible, since the gelatinous tissue of the mesoderm is apt to shrink if the object be left too long in aqueous fluids. The tissues were decalcified with very weak nitric or hydrochloric acid in 90 % alcohol. For maceration, the Hertwig's well-known mixture of osmic and acetic acid gives good results."

DR. ANDRES' METHODS OF TREATING ACTINIÆ.

Among the various methods employed by Andres in killing the Actiniæ, the three following, given in the order of their excellence, are said to have worked most satisfactorily : —

A. Corrosive sublimate.—With small animals, a hot solution, used in the manner recommended by Lang, gives good results; with larger animals, where this mode of treatment fails, the fluid must be injected. The cannula of a glass syringe, filled with the hot fluid, is inserted into the mouth at the mo-

---

[1] Mitth. a. d. Zoöl. Station zu Neapel, V., p. 3, 1884.

ment it opens, which act habitually follows on gently touching the lip. After injecting, the hot solution is poured into the glass containing the animal and a small quantity of sea-water.[1]

If the operation is cleverly performed, the animal remains fully expanded, as the mechanical pressure of the injected fluid prevents contraction.

After from five to fifteen minutes, the animal is washed in distilled water, and allowed to remain twelve hours in 50 % alcohol,[2] then passed through the higher grades of alcohol. Borax-carmine and hæmatoxylin used for staining.

B. Glycerine and Alcohol.[3] —

Glycerine	20 parts
Alcohol (70 %)	40 "
Sea water	40 "

This mixture, poured very slowly into the containing glass, often gives very good results, both for anatomical and histological purposes.

C. Nicotine and Tobacco Smoke. — *a.* A solution of nicotine (1 g.) in sea water (1 l.), conducted into the vessel containing the animal fully expanded in a half litre of sea-water, by means of a thread sufficiently large to empty the flask holding the nicotine solution in the course of twelve hours.

b. The vessel containing the animal in an extended condition, covered by a bell-jar in which tobacco smoke is confined, until the animal becomes completely benumbed.

After being deprived of sensibility by either of these methods, the creature may be killed in corrosive sublimate or in picro-sulphuric acid.

D. Andres finds that in the use of chloroform, dropped slowly into the water, or administered in form of vapor, maceration usually sets in before the power of contracting is lost. Good preparations of the internal parts may be obtained by injecting a weak solution of osmic acid. The method of freez-

---

[1] *Andres.* "Intorno all'Edwardsia Claparedii," in the Proceedings of the "Reale Accademia dei Lincei," Vol. V., Ser. 3, Mar. 7, 1880, p. 9.

[2] A little camphor (1:100) added to the alcohol will facilitate the removal of the sublimate.

[3] This method originated with Salvatore Lobianco.

ing has also been employed with some success. For this purpose, three vessels are placed one within the other, the central one containing the Actinia, the middle one ice and salt, and the outer one cotton.

The ice containing the congealed animal is dissolved in alcohol or an acid.

E. Maceration. — It is often important to see the cells of a tissue *in situ* before freeing them with needles. In such cases Dr. Andres proceeds as follows: —

1. Killed with corrosive sublimate.
2. Left in 25 % alcohol twenty-four hours.
3. Soaked for a short time in a very thin solution of *gum arabic*, then in a somewhat thicker solution, and finally imbedded in a very thick solution.
4. Hardened in 90 % alcohol.
5. Thick sections prepared for dissection with needles. The sections are placed on a slide in water, which dissolves the gum.

PREPARATION OF THE ECHINORHYNCHI.[1]

It is a very difficult matter to kill Echinorhynchi instantly. This cannot be done either with corrosive sublimate or strong osmic acid, even after preliminary treatment with tobacco smoke or chloroform. Thus treated, they contract strongly, and remain so after death.

Much the best results are obtained by killing gradually with $\frac{1}{10}$ % osmic acid, in which they contract during the first hours, but stretch out again and die fully extended. This method causes slight swelling, but does not seriously injure the object for histological investigation. In specimens left for twenty-four hours in the osmic acid, it is easy to isolate under the dissecting microscope the subcuticula, and the two layers of muscle fibres (circular and longitudinal). For the study of the internal organs, the Echinorhynchi should be cut open immediately after death and transferred to a $\frac{1}{100}$ % solution of osmic acid. The preservation of specimens thus treated may be accomplished in the following manner: after carefully

---

[1] A. Sæfftigen, Morph. Jahrb., x., Heft 1, pp. 120-163, 1884.

washing away the osmic acid, place the objects in a very dilute solution of potassic acetate in an open vessel, and leave them for two or three days, during which much of the solution evaporates. Finally transfer to a saturated solution in order to clarify so far as possible. Very beautiful preparations are said to be thus obtained.

The course of the nerves may be easily traced in specimens that have lain several days in 1% formic acid. The tissues swell up strongly and become quite transparent so that the nerves can be seen. If the muscular layers be separated from the subcuticula in specimens thus treated, and then stained in gold chloride, the lateral nerve trunks may be clearly shown. For the histological study of the nerves, the Echinorhynchi should be treated with chromic acid and then stained deeply with borax-carmine.

Chromic acid preparations are also best for the study of the subcuticula. Echinorhynchi live for days in a $\frac{1}{15}$% solution of chromic acid, but eventually die in a fully extended condition. Such preparations, after treatment with alcohol, may be colored at once; or, after washing a day or more in running water, exposed to the action of osmic acid, and then colored in borax-carmine.

For the study of the sexual organs, a very dilute picro-sulphuric acid (1 part of the acid to 8–10 parts of water) is recommended.

The tissues of the Echinorhynchi are not easily stained. Borax-carmine, which, according to Sæfftigen, is the best staining fluid, must be allowed to act a long time (often one or more days); after a deep stain has been taken, the preparation should be partially decolored by the use of hydrochloric acid in the ordinary way.

METHODS OF STUDYING THE SO-CALLED LIVER OF THE CRUSTACEA.[2]

For the study of fresh tissues, Dr. Frenzel places a small piece of the organ on the slide, *in the blood of the individual from which it was taken;* or, *in sea-water diluted until the*

---

[2] Johannes Frenzel. Mittheil. a. d. Zool. Station, v., p. 51, 1884.

salt contained amounts to about 1½–2% (1 part distilled water and 1 part sea-water from the Bay of Naples). The so-called "physiological salt solution" (¾%) worked unfavorably, causing maceration.

Various fluids were employed for killing and hardening, partly for determining the effect of different reagents on the nuclei and the protoplasm, and partly for finding the best means of preparing the object for sectioning.

Very good preparations were obtained with *warm* alcohol from 70% to 90%; while direct immersion in absolute alcohol did not prove advantageous. This treatment gave good results for the cell protoplasm, but destroyed the structure of the nuclei. Still better results were obtained for the cells (not for the nuclei) by adding a few drops of iodine to 70% alcohol.

The most satisfactory results were reached by immersing the object in a saturated aqueous solution of corrosive sublimate from 10–30 minutes, then washing with water, and finally replacing the water gradually with alcohol.

Perenyi's fluid gave best results when combined with corrosive sublimate. The object was left from 5–10 minutes in the first-named fluid, then transferred to the second and left for the same time.

While these methods were good for the Decapods, Amphipods, and Phronimidæ, the Isopods required a different treatment. With these, Kleinenberg's picro-sulphuric acid, diluted with an equal volume of water, and allowed to act 15–20 minutes, gave much better preparations than the sublimate solution.

A METHOD OF SILVER STAINING FOR MARINE OBJECTS.[1]

"The principle of this method was suggested to me by Dr. W. H. Ransom, and consists in the replacement of distilled water in the ordinary process of silver staining by a solution of a neutral salt not precipitable by silver nitrate, and of the same specific gravity as sea-water. *Loxosoma* and *Pedicellina* were the first objects investigated, and these animals are not killed

---

[1] S. F. Harmer. Mitth. d. zool. Station z. Neapel, v., p. 445, Nov., 1884.

by an exposure of as much as half an hour to a 5% solution of potassic nitrate in distilled water. It is thus quite easy to free the tissues from the greater part of their chlorides by washing with the above-mentioned solution of potassic nitrate; from this, the objects are transferred (naturally without the formation of any precipitate) for 4 or 5 minutes to a solution of silver nitrate ($\frac{1}{2}$–1%, according to circumstances). After reduction of the silver during exposure to light in the nitrate solution, the tissues may be mounted permanently either in glycerine or in Canada balsam. Very beautiful preparations of *Loxosoma* were easily obtained by the use of osmic acid and picro-carmine after treatment with silver nitrate. The animal may be transferred directly from the silver solution to osmic acid ($\frac{1}{2}$ %) and thence to picro-carmine, reduction taking place during the process, or the osmic acid may be added after the silver has been already reduced in the potassic nitrate. In successful preparations made in the above manner, the limits of all the cells of the epidermis and of the alimentary canal are exceedingly sharply marked out, the nuclei of these cells, as well as of the muscle cells, connective tissue corpuscles and other tissue elements being very distinctly stained. Besides the above-mentioned forms, I have obtained good results with the epidermis of *Medusæ*, *Hydroids*, *Saxitta* and *Appendicularia* (tail).

"Amongst sponges, Dr. G. C. J. Vosmaer has, by means of the above method, easily demonstrated an outer epithelium in *Chondrosia*, where F. E. Shulze first has been unable to detect a cell layer of this kind even by the use of silver nitrate;[1] in *Thenea* where Sollas[2] has observed no cell outlines, and in many other sponges. Dr. Ed. Meyer has successfully applied the same method to the epidermis and peritoneal epithelium of Annelids (*Tomopteris*, *Amphictenidae*), to *teleostean* ova, etc. He has obtained good nuclear coloration by transferring the tissue after reduction in potassic nitrate to alcohol, and subsequently staining with Paul Mayer's alcoholic carmine.[3] In

---

[1] Zeitschr. f. wiss. Zool. XXIX., p. 107, and XXXI., p. 290.
[2] Ann. and Mag. of Nat. Hist., 5th Series, IX., p. 445.
[3] Mitth. d. Zoolog. Stat. zu Neapel. IV., p. 521.

Brachiopoda, Dr. J. F. Van Bemmelin has had no difficulty in using the method for the investigation of the epidermis and peritoneal epithelium.

"Few animals seem to resist the action of potassic nitrate to so great an extent as *Loxosoma* and *Pedicellina*, most forms being either immediately, or after a few minutes, killed by an immersion in a 5% solution of this substance. Even in many of these cases, the tissues suffer very little histological change, and can be easily stained by silver nitrate. It is possible that many other salts may be used more advantageously than potassic nitrate in washing the chlorides from the tissues without killing the animal. A $4\frac{1}{2}$% solution of sodic sulphate may be used instead of the potassic nitrate, over which, however, in most cases it has no obvious advantages.

"R. Hertwig[1] has described a method for the silver staining of marine animals, consisting in treating with dilute osmic acid, washing with distilled water, until the chlorides are removed, subsequently placing in a silver nitrate solution, and reducing in distilled water. By this method, Hertwig has been able to obtain sharply defined cell limits in the ectoderm of *Ctenophora*, but not over the whole surface of the epithelium. I have myself employed Hertwig's method in *Loxosoma* and *Pedicellina*, but with results not nearly so satisfactory as in the method I have described, the limits of the epidermic cells being not visible over the whole surface, whilst after the employment of Dr. Ransom's method in these two genera, I have invariably found that every epidermic cell was distinctly marked out by a sharp contour."

THE DEMONSTRATION OF CELLS IN THE CORPUS VITREUM OF CYPRINOID FISHES.

The vitreous body of the eye of the fish is, in many species, according to H. Virchow,[2] entirely destitute of cells; while in others its whole surface is covered with them. The form of these cells varies for different species, the peculiari-

---

[1] Jenaische Zeitschr. XIV., p. 324.
[2] Arch. f. mik. Anat., XXIV., H. 2, pp. 106–108, Oct., 1884.

tics being sufficiently well marked to serve as a means of distinguishing groups, families, and even genera and species.

In order to study the vessels and cells of the vitreous humor, it is necessary to make a preparation of a portion of the peripheral zone ("Grenzhaut"), free from surrounding tissues. Place the eye, freed from the sclerotic and choroid investments, (1) in a 1 % solution of corrosive sublimate, heated to about 30° C., and leave it here 7 hours, the solution meanwhile being allowed to cool; or (2) in Müller's fluid for 24 hours. Either method of treatment will secure, in most cases, the separation of the retina from the vitreous body, which is the first important step in the preparation. The advantages of the first method are that it hardens just enough to render further preparation easy, and preserves well the color of the tissues; but it produces shrinkage in the nuclei. Müller's fluid is a better preservative for the nuclei, but hardens less, and injures the clearness of the color.

The process of staining will be most successful if undertaken immediately after washing in distilled water — before hardening in alcohol. For this purpose the preparation should be left in a weak solution of hæmatoxylin 24 hours, washed for an hour or more in a $\frac{1}{2}$ % solution of alum, and then placed in Ogatha's solution of eosin 12 hours (eosin + 60 alcohol + 140 water.)[1]

After the preparation is stained, it may be washed in absolute alcohol, and then examined in glycerine (?).

Good preparations for showing the distribution of the cells and blood vessels may be obtained in the following manner: —

After treatment with corrosive sublimate, cut off the hinder portion of the preparation, and remove the retina; then, by the aid of two pairs of forceps, separate a piece of the superficial layer of the vitreous body from the remaining portion, and place it (external side down) on a slide; after placing the cover-glass, press it lightly, and then put slide and all into a dish of alcohol. After some hours, the cover-glass, with the preparation adhering to it, may be loosened from the slide, and the remaining steps of staining, washing, anhydrating, clarifying, and mounting in balsam accomplished without difficulty.

---

[1] Arch. f. Anat. u. Phys. Abth., 1883, p. 409.

THE MUCOUS MEMBRANE OF THE STOMACH.

In dealing with this subject, Nikolai Trinkler[1] has improved considerably the methods of preparation hitherto employed.

1. As soon as the animal is dead, the stomach, with a portion of the œsophagus and duodenum, is cut out very quickly, and opened along the large curvature. The surface of the mucous membrane is first tested in different places with litmus paper, and then washed with a stream of distilled water.

2. In animals with a thick-walled stomach, the external muscular layer is separated, and then the mucous membrane cut into small pieces, some of which are placed in hardening fluids, while others are placed in some indifferent fluid (humor aquæus, iodized serum, $\frac{1}{2}$ % NaCl) for examination in fresh condition. In order to prevent putrefaction, in the latter case, they may be kept in an atmosphere of water vapor and carbolic acid.

For Hardening. — Müller's fluid (1 or 2 days) is the best agent. Preparations hardened in chromic ($\frac{1}{2}$–1 %) or picric acid should not be transferred directly to alcohol, but first left for a day or more in Müller's fluid, then treated with alcohol of successively higher grades.

Preparations with osmic acid ($\frac{1}{10}$–$\frac{1}{2}$ %) should be washed a long time in distilled water, then placed in a saturated solution of potassic acetate, and finally examined in glycerine.

For Maceration of the epithelium and glandular elements, Müller's fluid was used, either alone or in combination with a weak solution of NaCl, as employed by Kutschin. Alcohol ($33\frac{1}{3}$ %), chloral hydrate (5 %), chromate of ammonia, salt solution (35 %), and a mixture of 1 vol. chromic acid ($\frac{1}{80}$ %), 1 vol. chloral hydrate (5 %), and a few drops of acetic acid, were also employed for the same purpose. This mixture and that of Kutschin give excellent results.

In Staining, Heidenhain's method[2] of using weak dyes and allowing them to act a long time may be followed, as a general rule. Litmus and chlorophyll were also used as dyes. A

---

[1] Arch. f. mikr. Anat., XXIV., Oct., 1884, pp. 174–176.
[2] Arch. f. mikr. Anat., VI., p. 402.

strong solution of litmus, of absolutely neutral reaction, must be employed, and this serves both as a staining and as a microchemical agent. Chlorophyll was obtained by soaking the leaves of Syringa vulgaris in strong alcohol 24 hours; the dark green fluid thus obtained was filtered, and then ·evaporated. The dry residue was dissolved in a little distilled water, and again filtered.

Following in the line of Heidenhain and Rollett, the different functional conditions of the glands, induced by feeding, starving, etc., were investigated; and more extreme conditions were reached by feeding with phosphor and alcohol.

THE CONTINUITY OF PROTOPLASM IN VEGETABLE STRUCTURES.

W. Gardiner [1] gives the following methods of detecting protoplasmic threads, which pass from cell to cell: —

Chlor-Zinc-Iod Method. — 1. Sections of endosperm are first stained with iodine and mounted in chlor-zinc-iod, in which they remain about 12 hours.

2. After washing well, stain with picric Hoffmann's blue (prepared by dissolving the dye in 50 % alcohol, which has been saturated with picric acid).

3. Wash again, and mount in glycerine; or, after treating with the usual grades of alcohol, and with clove oil, mount in balsam.

In cases where the tissue swells quickly under the action of chlor-zinc-iod, the exposure need not be so prolonged; and excessive swelling must be prevented by the use of alcoholic iodine at the outset, and by washing with alcohol instead of water.

As to the time of action, the manipulation of the reagents must be varied to suit the requirements of different kinds of tissue.

Molybdic Acid Test. — If a section of some living endosperm is treated with a solution of molybdic acid in strong sulphuric acid, the cell-wall will swell up, and the threads which traverse it will soon assume a blue color, while the main mass of protoplasm becomes intensely blue. The cell-wall itself remains uncolored. This very delicate reaction demonstrates the protoplasmic nature of the threads.

---

[1] Arbeit. Bot. Inst. Würzburg, III., pp. 53–60, 1884. Cf. Journ. Roy. Micr. Soc., IV., Part. 5, pp. 637–641, 1884.

CHAPTER XIII.

RECONSTRUCTION FROM SECTIONS.

ORIENTATION IN MICROTOMIC SECTIONS.

IF an object has been cut into serial sections and mounted, the value of the series for microscopical investigation will depend not only on the success with which each step in the preparation has been attended, but also on our ability to grasp all the topographical relations of each section. It is not enough to know the *region* through which a section passes; we must have the means of ascertaining to within a very small fraction of a millimeter the exact path of the knife. Such precise orientation can only be arrived at in an indirect way; but the improved instruments and methods of section-cutting make its attainment a no very difficult task. To determine the *locus* of sections with accuracy, several conditions must be fulfilled. The sections must be made of *uniform thickness*, arranged in *serial order*, and all *similarly disposed*. With these conditions satisfied, the *plane of section* determined, and an accurate surface view of the object obtained prior to imbedding, it becomes an extremely simple matter to know what portion of the surface view is represented by any given section. The following data will furnish an illustration:—

Blastoderm of the chick, 5^{mm} long.
Surface view magnified 20 diameters.
Thickness of each section $.05^{mm}$.
Plane of section at right angles to the long axis of the blastoderm.

From these data we know that there should be just 100 sections, and that each section must correspond to 1^{mm} of the surface view.

Now if we draw a line at one side of the surface view, parallel to, and of equal length with, its long axis, and divide this line into 100 equal parts, the number of the section will correspond to the same number on the scale, and the exact position of the section be recognized at a glance.

Of the conditions above named as essential to an exact knowledge of the *locus* of any given section, the only one likely to present any serious difficulties is that of obtaining sections of uniform thickness. Where the object-carrier and vernier are combined, and moved directly by the hand, it is extremely difficult, if not impossible, to obtain that degree of uniformity required for exact topographical study. In the best microtomes now in use, the carrier is moved only indirectly by the hand, through a micrometer screw, and its movements are thus brought under perfect control.

THE RECONSTRUCTION OF OBJECTS FROM SECTIONS.

The importance of attending to all available means of orientation will be best understood by those who know how to make use of sections in the reconstruction of objects or parts of objects. Suppose the only material at the disposal of an investigator to be a single small object, and that the rarity of the object renders its replacement extremely improbable. How shall the object be treated in order that the most exhaustive knowledge of all the details of its inner structure may be obtained? One might be tempted to lock it up as a cabinet rarity, if he did not know how to make a single series of sections tell the whole story. If the preliminary steps have been correctly taken, it is possible to construct from serial transverse sections a median sagittal (longitudinal and vertical), or frontal section, or a section in any desired plane. From the same series may be constructed also surface views of internal organs, which are inaccessible to, or unmanageable by, any of the ordinary methods of dissection.

It frequently happens that sections can be obtained by construction that could not be obtained by any direct means. For example, we may desire a frontal section of a vertebrate embryo that will show all the parts that lie in the same level with the chorda, or a sagittal section that will represent a median plane. It is evident that no such sections can be directly obtained, owing to the axial curvature of the embryo; but they can easily be constructed from transverse sections. It is here that we see some of the great advantages to be derived from the use of the microtome. It not only overcomes the opacity of objects, but it also enables us to represent curved and twisted surfaces in plane surfaces. The ability to construct sections at right angles to the actual planes of section is the key to the next and final step, — "the plastic synthesis" of the sectioned object.

METHOD OF RECONSTRUCTION.

Professor His was the first to make known a method of reconstruction.[1] Others have since made use of the same method for different purposes. A. Seessel, a former pupil of Professor His, employed it in a work on the development of the fore-gut.[2] Rosenberg made use of it in the construction of frontal views of the sacrum;[3] and Krieger, in the investigation of the central nerve-system of the crayfish.[4] The method is well illustrated by two figures (11 and 12, Pl. XXXI.), given by Krieger; and these figures are well worth examination, as they show how to proceed when the plane of section is not quite at right angles to the axis of the object. Professor His has also constructed frontal and profile (sagittal) views of the human embryo by the same method, and has explained the process in Part I., p. 10, of his "Anatomie menschlichen Embryonen."

For an illustration, we will take the data given under the head of orientation, and indicate how a surface view could be constructed from a series of transverse sections of the germinal disc of the chick. We should first draw 100 parallel zones on

---

[1] His. "Untersuchungen ü. d. erste Anlage des Wirbelthierleibes," p. 182, 1868. "Neue Untersuchungen ü. d. Bildung des Hühnerembryo, in Arch. f. Anat. u. Physiol., anat. Abth.," p. 122, 1877.

[2] Seessel. "Arch. f. Anat. u. Physiol., anat., Abth.," p. 449, 1877.

[3] Rosenberg. Morph. Jahrb. Vol. I., p. 108, 1875.

[4] Krieger. Zeitschrift f. wiss. Zool. Vol. XXXIII., p. 531, 1880, and Zool. Anzeiger, p. 369, 1878.

a sheet of paper, each zone corresponding in thickness to a single section (1^{mm}).

A median line would then be drawn at right angles to these zones; this line would represent the length of the disc magnified 20 diameters (100^{mm}). We should next make an outline drawing of the first section enlarged the same number of diameters as before. The width of this drawing and its parts (primitive streak, embryonic rim, etc.), could then be indicated in the first zone by dots placed at the proper distance on the right and left side of the median line. The dots for each succeeding section having been placed in their corresponding zones, nothing further would remain to be done, except to connect the dots of corresponding parts in the several zones, and shade according to the requirements of the case.

If the plane of section is not quite perpendicular to the axis of the object, one has only to determine the angle which the axis makes with the plane of section, and draw the median line so that it forms the same angle with the parallel zones. Such a case has been clearly illustrated by Krieger.

In the construction of sagittal sections, a profile line (dorsal line, etc.), will serve as the ground line.

BORN'S METHOD OF RECONSTRUCTING OBJECTS FROM MICROSCOPIC SECTIONS.[1]

Dr. G. Born describes in detail a very ingenious method of constructing models of objects from serial sections. By the aid of the camera, the outlines of the sections are transferred to wax plates, which are then cut out so as to correspond in outlines as well as dimensions to the sections equally magnified in all three directions. With plates thus prepared, it is only necessary to put them together in the proper order to obtain a complete model. The method is simple and extremely useful, especially in investigating objects with complex internal cavities.

Born has made use of the method in studying different parts

---

[1] G. Born. "Die Plattenmodellirmethode." Archiv f. mik. Anat. XXII., p. 584, 1883. First described in Morphol. Jahrbuch, II., p. 578, 1876.

of the vertebrate head; Swirski, in elucidating the development of the shoulder-girdle of the pike; Stöhr, in tracing the development of the skull of Amphibia and Teleostei; and Uskow in studying the development of the body cavity, the diaphragm, etc.

An Illustration of the Method. — Born makes use of three rectangular tin boxes of equal size, each measuring $270^{mm} \times 230^{mm} \times 2\frac{1}{2}^{mm}$. Sections should be made about $\frac{1}{25}^{mm}$ thick (never thinner than $\frac{1}{50}^{mm}$). If we desire to construct a model of an object from serial sections $\frac{1}{30}^{mm}$ thick, which shall be magnified 60 diameters, then the wax plates must be made 60 times as thick as the sections, *i.e.*, 2^{mm} thick.

The surface of a plate that could be made in a box of the above named dimensions contains $62.100 \square^{mm}$; and the volume of such a plate 2^{mm} thick would therefore be 124.2^{ccm}. The specific gravity of common raw beeswax amounts to .96–.97. For use, it requires only to be melted and a little turpentine added to make it more flexible. Thus prepared, its specific gravity is about .95; and this number has been found sufficiently accurate in all cases. The weight of the wax required to make one plate of the above size, will, accordingly, be 117.99 gr., or, in round numbers, 118 gr. The wax having been weighed and melted, the tin box is first filled $1\frac{1}{2}^{cm}$ deep with boiling water, and then the melted wax poured upon the water. If the water and the wax are quite hot, the wax will generally spread evenly over the surface; if gaps remain, they can be filled out by the aid of a glass slide drawn over the wax. As soon as the plate has stiffened, and while it is still soft, it is well to cut it free from the walls of the tin box, as further cooling of the water and the box might cause it to split. By the time the water becomes tepid, the plate can be removed from the water to some flat support, and left till completely stiffened. Half a hundred plates may thus be prepared in the course of a few hours.

The outlines of the section are transferred to the plate in the following manner: A piece of blue paper is placed on the plate with the blue side turned toward the wax, and above this is placed a sheet of ordinary drawing paper. The outlines are drawn on the latter by the aid of a camera, and, at the same time, blue outlines are traced on the wax plate. The plate can then be laid on soft wood and cut out by the aid of a small knife. Thus a drawing and a model of each section are prepared. The plates thus prepared can be put together in the proper order, and fastened by the aid of a hot spatula applied to the edges.

APPENDIX.

APPENDIX.

APPENDIX.

METHODS OF INJECTION.

THE PREPARATION OF DRY INJECTION MASSES.[1]

The variously colored gelatine emulsions in common use as injections keep for only a short time, and have therefore to be prepared as occasion arises for their use. The dry emulsions recommended by Fol are very easily prepared and convenient in use. As they will keep for any length of time, they can be prepared in quantities, and thus be ready for use at any moment.

Carmine Emulsion. — One kilogram gelatine (softer kind used in photography) soaked in water for a few hours until thoroughly softened; after turning off the water, heat the gelatine over a water-bath until liquified; and then add to it, little by little, one litre of a strong solution of carmine in ammonia.[2] The mixture, stiffened by cooling, is cut up, and the pieces packed in a fine piece of netting. Vigorous pressure with the hand under water forces the emulsion through the net in the form of fine strings or vermicelli. These strings are placed in a sieve, and washed until they are free from acid or excess of ammonia, then collected, and re-dissolved by heating. The liquid is poured upon large sheets of parchment which have been saturated with paraffine, and these sheets are then hung up to dry in an airy place. The dried layers of the emulsion, which are easily separated from the parchment, may be cut into

---

[1] Fol. Beiträge zur histologischen Technik. Zeitschr. f. wiss. Zool., XXXVIII., p. 491, 1883.

[2] Cf. Lehrbuch d. vergl. mik. Anat., p. 13, when a somewhat modified process is given.

strips, and placed where they are protected from dust and dampness.

The carmine solution used in this emulsion is prepared as follows: —

A strong solution of ammonia is diluted with 3-4 volumes of water, and carmine added in excess. After filtering, the solution is mixed with the gelatine, and then enough acetic acid added to change the dark purple-red into blood-red. It is not necessary to completely neutralize the ammonia.

The dry emulsion requires only to be placed in water for a few minutes, and melted over the water-bath, to be ready for use.

Blue Emulsion. — A slightly modified form of Thiersch's formula. —

1. To 300^{ccm} of melted gelatine add 120^{ccm} of a cold saturated solution of green vitriol (ferro-sulphate).

2. To 600^{ccm} of melted gelatine add first 240^{ccm} of a saturated solution of oxalic acid, then 240^{ccm} of a cold saturated solution of red prussiate of potash (potassic ferricyanide).

3. No. 1 poured slowly into No. 2 while stirring vigorously; the mixture heated for fifteen minutes.

4. After cooling, the emulsion is pressed through netting, the vermicelli washed and spread on waxed paper for drying. In this case the vermicelli must be dried directly, as they do not melt well without the addition of oxalic acid.

The dry vermicelli are prepared for use by first soaking in cold water, and then heating with the addition of oxalic acid enough to reduce them to a liquid.

Black Emulsion. — 1. Soak 500 g. gelatine in two litres of water in which 140 g. of common salt have previously been dissolved, and melt the mass on the water-bath.

2. Dissolve 300 g. nitrate of silver in 1 litre distilled water.

3. No. 2 poured very slowly into No. 1 while stirring. An extremely fine-grained emulsion may be obtained by using 3-4 times as much water in Nos. 1 and 2.

4. No. 3 pressed into vermicelli, as above, and then mixed with No. 5 by clear daylight.

5. Mix 1½ litres cold saturated potassic oxalate with 500^{ccm} of a cold saturated solution of ferro-sulphate.

6. No. 4 mixed with No. 5 gives a thoroughly black emulsion, which should be washed several hours, again melted, and finally poured in a thin layer on waxed paper.

A gray-black emulsion may be obtained by using 240 g. potassic bromide in the place of common salt in No. 1, the remaining operations being the same.

A STARCH INJECTION MASS.[1]

A coarse injection mass, which is cold-flowing, may be forced nearly to the capillaries, rapidly hardens after injection, leaves the vessels flexible, does not dull dissecting instruments, is suitable for permanent dry or alcoholic preparations, is simple in its manipulation, cleanly and economical, seems to be fully realized in the starch mass introduced by Ad. Pansch, of Kiel, and since recommended, with various modifications, by Wikszemski, Dalla Rossa, Meyer, and Browning.[2]

As starch is insoluble in alcohol and cold water, it becomes hard when injected into the blood vessels simply by the exudation of the liquid with which it is mixed. (That the starch grains forming the mass remain entirely unchanged may be easily demonstrated by making a microscopic examination of the contents of an injected vessel.)

The mass originally recommended by Pansch consisted of wheat flour and cold water, to which was added a sufficient quantity of the desired coloring matter. Later experiments have shown that pure starch is better than flour.

Mass for Ordinary Injections. —

Dry starch ("laundry" is good)	1 vol.
2½ % aqueous solution of chloral hydrate	1 "
95 % alcohol[3]	½ "
Color	½ "

---

[1] S. H. Gage. New York Medical Journal, June, 1884.

[2] See Ad. Pansch, "Archiv für Anatomie und Entwickl.," 1877, pp. 480-482, and 1881, pp. 76-78; Wikszemski, same Journal, 1880, pp. 232-234; Dalla Rossa, same, pp. 371-377; Herm. von Meyer, same, 1882, pp. 60, 61, and 1883, pp. 265, 266; Browning, "Annals of Anatomy and Surgery," 1884, pp. 24, 25.

[3] The chloral and alcohol prevent fermentation in the mass when it is kept in stock; the alcohol also increases the fluidity, and likewise the more rapid hardening in the vessels; both, of course, act as a preservative upon the animal injected.

Since almost any animal injected may afford some organ worth preserving, it seems better to employ permanent colors for tinging the mass. Among those which are available, probably vermilion, red lead, ultramarine, chrome orange, yellow, or green are preferable.

Preparation of the Color.—

Dry color	1 vol.
Glycerine	1 "
95 % alcohol	1 "

To avoid lumps, which would clog the cannulæ, or small vessels, the color is thoroughly ground with the liquid in a mortar. It is stored in a well-stoppered bottle, and is prepared for use simply by shaking.

Special Mass.— For the injection of brains, and, perhaps, for other rapidly perishing specimens, it seems best, as suggested by Professor Wilder, to use strong preservatives in preparing the mass:—

Corn starch (that used for food)	1 vol.
5 % aqueous solution of chloral hydrate	$\frac{1}{4}$ "
95 % alcohol	$\frac{1}{2}$ "
Color	$\frac{1}{4}$ "

For convenience and economy, a considerable quantity of either of the masses described above may be prepared at once, and kept in a wide-mouthed specimen or fruit jar. A smooth stick in each jar is convenient for stirring the mass, which should always be done just before using. The syringe may be filled directly from the jar, and any mass remaining in the syringe after the injection is finished may be returned to the jar.

If it is desired to have the mass enter very fine vessels, some of the stock mass, as given above, diluted with an equal volume of water or chloral solution, may be injected first, and immediately followed by the undiluted mass, or, for large animals, a mass containing twice the usual amount of starch. In whatever form the starch is used, it is necessary to work somewhat expeditiously, because the exudation of the liquid in the smaller vessels takes place so rapidly that the mass hardens very quickly in them. The larger the vessel, the more slowly, of course, do the exudation and, consequently, the hardening, take place. It sometimes happens that large vessels, like the aorta, are not fully distended after the exudation of the liquid. In this case some mass containing double the ordinary amount of starch can be advantageously injected in two hours, or longer, after the first injection.

Dry Preparations. — Finally, if vessels injected with the starch mass are dissected free, soaked a day or two in Wickersheimer's preservative, and then dried, they retain their form, and, to a great degree, their flexibility.

CELLOIDIN INJECTIONS.

In the formation of injection masses, collodion plays still another important role, for the discovery of which we are indebted to Schiefferdecker.[1] It can be made to penetrate easily very fine blood-vessels; and its viscosity protects them against injury. Its application is simple and easy, and in all these respects it is said to be superior to the masses hitherto employed for "corrosion preparations." It has a slight shrinkage, but not enough to form a serious drawback. It is prepared in different ways, according to the color to be given to the injection.

A. Asphalt-Celloidin Injection. — 1. Pulverized asphalt placed in a well-closed bottle of ether, and allowed to remain 24 hours, during which the mixture must be several times shaken.

2. The brown-colored ether is turned off, and small pieces of celloidin dissolved in it until the solution flows like a thick oil.[2]

B. Vesuvian-Celloidin Injection. — 1. Make a saturated solution of Vesuvian in absolute alcohol.

2. Dissolve in this pieces of celloidin until the desired consistency is reached.

The brown injection thus obtained is less satisfactory than that formed from asphalt, as its color fades somewhat.

C. Opaque Celloidin Injections. — 1. Dissolve celloidin in absolute alcohol and ether in equal parts.

2. Add vermilion or Prussian blue to color.

---

[1] l. c., p. 201.

[2] The pulverized asphalt can be used many times over for coloring the ether, as very little of it will dissolve in 24 hours.

The coloring substance should be mixed with a small quantity of absolute alcohol, and then reduced to great fineness by continued trituration in a mortar. To the thick, paste-like mass thus obtained, the solution of celloidin is next added. The amount of coloring substance should be as little as possible, as the mass will otherwise be too brittle. If a fine injection is required, the mass should be filtered through flannel moistened with ether.

The syringe employed must be entirely free from fatty substances, as these render the injection-mass brittle. If the piston does not fit the syringe-tube sufficiently closely, it may be wound with a little gauze. The cannula should be filled with ether before it is inserted, and tied in place, and again filled just before it is joined to the syringe.

In using a mass dissolved in alcohol and ether, it is well to add a little ether, which will spread over the surface, and thus prevent the formation of a film. The injection should be made moderately quickly, as the mass stiffens soon after contact with the tissues. After injection, the syringe and cannula should be cleaned with ether.

The injected organ is placed in hydrochloric acid, diluted more or less, according to the danger of shrinkage. It is left in the acid, which is occasionally renewed, until the tissues are sufficiently corroded to be easily washed away by a slow and steady stream of water, conducted through rubber tubing connected with a water pipe. The preparation may then be left in water for some days or weeks, in order to free it from remaining fragments of tissue by gradual maceration. The preparation, when finished, may be preserved either in glycerine or a mixture of glycerine, alcohol, and water in equal parts.

The asphalt-celloidin mass is the one most highly recommended by Schiefferdecker.

OSBORNE'S METHOD OF INJECTING THE ARTERIES AND VEINS.

In this method two injecting fluids are employed, the first having a density that will allow it to pass the capillaries easily, while the second is of such a density that it will be arrested at the capillaries. The whole vascular system may thus be injected from the arterial bulb.

The method is employed in the following manner:—

1. The animal to be injected is immersed in tepid water, and the heart laid bare.

2. The apex of the single ventricle, in the case of an amphibian, or of the left ventricle, in the case of higher animals, is then widely opened, and the blood allowed to flow from the auriculo-ventricular aperture.

3. The cannula is next inserted so that it reaches into the arterial bulb, and two ligatures made at the points indicated by 1st L and 2d L.

4. When the body is thoroughly warmed, an ordinary red or purple gelatine mass is slowly injected. The second ligature having been left loose, a quantity of blood, gradually followed by the injecting mass, flows from the auriculo-ventricular opening.

5. As soon as the gelatine mass runs quite clear, the second ligature is fastened, and the syringe replaced by another containing a red plaster-of-Paris injecting mass. The latter drives the gelatine mass contained in the arteries before it, as far as the capillaries. When the gelatine is well cooled, the animal is ready for dissection.

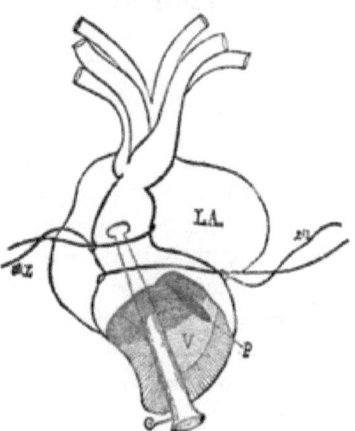

FIG. 36.—Illustrating method of preparing the frog's heart. V, ventricle; LA, left auricle; p, auriculo-ventricular opening; 1st L and 2d L, first and second ligatures; C, cannula.

The preparation may be well preserved for an indefinite time in Wickersheimer's fluid. Alcohol shrinks the gelatine, and thus produces breaks in the veins.

HAY'S METHOD OF MAKING DOUBLE INJECTIONS.

A. For Dissecting Purposes.—A cannula is fitted into the aorta of a cat, and a gelatine mass colored with carmine is injected until it is seen to flow from the right side of the heart; then, the tube conveying the red mass being detached, a tube conveying a blue gelatine mass is slipped over the same cannula, and the pressure again applied. Into this blue mass is mixed thoroughly a quantity of starch — preferably from wheat. This starch-bearing mass pushes the carmine mass before it until the starch grains enter the capillaries, and effectually plug them up. The arteries are thus left blue and the veins red. The first mass injected need not be unusually thin.

B. For Histological Purposes. — The capacity of the capillaries is so great, as compared with that of the arteries, that any commingling of the two colors is concealed in them. Carmine is used for the veins because of the ease with which it may be prepared, its permanence, and the facility with which it passes through the capillaries. On the other hand, the gelatine for the arteries may be colored with the coarser pigments, such as Prussian blue or ultramarine. The latter furnishes a beautiful blue. Vermilion is not suitable for the first injected mass, since, on account of its high specific gravity, it readily sinks to the lowest side of the vessels, drags behind, and causes a commingling of the colors. An additional reason for filling the veins with red rather than with blue is found in the agreeable and natural color given to the preparation.

Of course, a mass of plaster of Paris, injected after a gelatine mass, will drive it on until the plaster reaches the smallest vessels, thus producing a double injection. The starch mass recently proposed as a filling for blood-vessels will readily lend itself to the production of a double injection according to the method detailed above.

So far as I am aware, the usual method of producing a double injection of the blood-vessels, preparatory to making sections for the microscope, is to inject first a gelatine mass of one color into the artery until the increasing pressure gives notice that the mass is entering the capillaries, and immediately after to inject a differently colored mass into the vein. The injection being thus accomplished, one of two things, it seems to me, is likely to happen: either the vessels will not be well filled, or the mass intended for one set of vessels will be driven through into the other. To avoid these accidents, I have practised the method of filling both sets of vessels at the same moment and under exactly the same pressure.

The pressure is kept low at the beginning, so that all the arteries and veins shall be thoroughly filled before either mass begins to enter the capillaries. Then, as the pressure is increased, the differently colored masses meet each other in the capillaries, and, if the pressure on each is equal, the vessels may be filled as full as compatible with safety without danger of either color being driven from one set of vessels into the other. The way in which this result is accomplished will be understood better by reference to the accompanying drawing (Fig. 37). The desired pressure is secured by allowing a stream of water from a hydrant or from an elevated cistern to flow into a tight vessel. A two-gallon petroleum can does quite well. As the water flows in, the air is forced out through a rubber tube, A, into the wide-mouthed bottle, F, whose tightly fitting cork gives passage to two other glass tubes. These extend below, just through the cork, and above, connect respectively with the rubber tubes C and D. Into the side of F, near the bottom, is fitted another tube, E, reaching to a height of ten inches or more, open above and graduated into inches. If preferred, this tube may also pass through the cork and extend down well into the mercury with which F is partly filled. B is a bottle of suitable size, in which is contained a blue injection mass for filling the veins, and R, a similar bottle containing a red mass for the arteries. The interiors of these bottles are connected

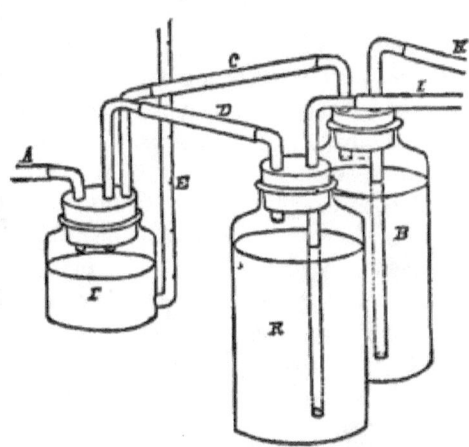

FIG. 37.— Apparatus for making double injections.

with the bottle F by the tubes D and C. Each of the bottles B and R has a tube, which, starting from near the bottom, passes through the cork, and is, a little above this, bent at right angles. With these are connected the rubber tubes, H and I. Now when water is allowed to flow into the reservoir mentioned above, the air is forced out through A into F, and thence along the tubes D and C into B and R. As soon as the pressure in these bottles becomes sufficiently great, the liquid which they contain will be driven out through the tubes H and I. If there should be any obstacle to the escape of these fluid masses, the pressure in all the vessels will rise and be registered by the height of the mercury in E.

If now it is desired to inject, for instance, the kidney of a pig, a cannula made of a glass tube must be fitted securely into the renal artery, and a similar one into the renal vein. The cannulæ must be of such a size that the rubber tubes H and I will fit them well. Heat the gelatine masses in the bottles B and R to the proper temperature, and keep them so heated until the injection has been finished. Special care must be taken with the tubes H and I to prevent the gelatine passing through them from becoming frozen. Now, having clamped the tube H, have an assistant turn on a small stream of water until the gelatine begins to flow slowly from I. If the diameter of the cannula is not too small, it may be held with the free end directed upward and filled with gelatine allowed to drop from the mouth of I. Then slip I over the cannula, unclamp the tube H, and when the gelatine from B has begun to flow, slip it over the cannula inserted in the vein. Then increase the pressure gradually until it has reached as high a point as experience has taught to be safe for the organ operated on.

By means of this apparatus, which will require the expenditure of only a few cents and a little ingenuity, double injections may easily be made of any organs whose veins are not provided with valves. I have made injections of the kidney, whose arteries and glomeruli became uniformly filled with the red mass, and whose veins and the system of capillaries surrounding the renal tubules became filled with the blue. The lungs and the liver are easily and successfully injected. I have been less successful in injecting **the** organs **that send away** their blood current through the portal vein; **but I have no** doubt that they too may be injected.

Triple injections of the liver may **be made by first injecting the** hepatic artery with **a green** mass **until the whole liver** assumes a green tint, and **afterwards injecting the portal vein** and the hepatic **vein with red and blue as above directed.**

The same apparatus **may be employed to make either single** injections or the **double injection described under the second** head of this paper, **by simply clamping one of the tubes C or D.** As a matter of course, care must be taken that all the corks fit tightly in the bottles, otherwise the internal pressure may force them out at the very moment when an accident will do the most damage.

MUSEUM METHODS.

ANÆSTHETIC AGENTS.

Carbonic Acid.—Fol[1] has made the important discovery that Cœlenterates and Echinoderms may be rendered insensible, and kept so for hours and even days, without injury, by saturating the water with carbonic acid. The containing vessel must, of course, be hermetically closed. The animal at once becomes insensible and motionless, but preserves its natural appearance, and recovers at once when again placed in pure sea-water. This method may be used, not only for obtaining life-like photographs, but also, as Dr. Fol suggests, for transporting animals alive. Fishes and molluscs do not survive this treatment, and crustaceans for only a short time.

Fol tried various narcotics, but found that small doses would not bring the animals to rest, while large doses acted as poisons. The same proved true of tobacco smoke and aqueous solutions of ether, chloroform, and ethyl bromide. Sulphydric acid and carbonic oxide gave satisfactory results in only a few cases.

Chlorhdyrate of Cocaine.—In order to kill in an ex-

---

[1] Zoologischer Anzeiger, No. 128, p. 698, 1882.

panded condition such animals as the fresh-water Hydra, Bryozoa, etc., it is only necessary to treat them in the manner recommended by G. J. Richard.[1] For example, place a colony of Bryozoa in a watch-glass with about 5^{cc} of water. When they are well expanded, add slowly $\frac{1}{2}^{cc}$ of a 1 % solution of chlorhydrate of cocaine. After five minutes, another like quantity of the solution is added. At the end of ten minutes more, the animals are perfectly insensible, and may be treated with the ordinary hardening reagents without danger of contraction.

WICKERSHEIMER'S FLUID.[2]

(1) 3000 g. boiling water; in this are dissolved: —
(2) 100 g. alum.
(3) 25 g. salt.
(4) 12 g. potassic nitrate (KNO_3).
(5) 60 g. potash (KHO).
(6) 10 g. arsenic trioxid (As_2O_3). To 10 litres of this mixture, after it is filtered and cooled, are added

(1) 4 litres glycerine.
(2) 1 litre methyl alcohol.

Dried Specimens. — If specimens are to be later preserved in a dried condition, they are allowed to soak in the fluid 6 to 12 days, according to their size, then taken out, and dried in the air. Muscles, tendons, etc., remain thus soft and pliable.

Hollow organs (lungs, intestines, etc.) are first filled with the fluid, before being placed in it.

To Preserve Color. — Specimens must not be dried, but kept in the fluid.

Preservation of Bodies for Dissection. — Injection of the fluid suffices to keep corpses for years. They will lose their fresh appearance, and the epidermis will brown. This can be prevented by rubbing the fluid over the surface and then preserving in air-tight vessels.

---

[1] Zool. Anzeiger, VIII., No. 196, p. 332, 1885.
[2] Zool. Anzeiger, Vol. II., No. 45, p. 669, 1879.

KOCH'S METHOD OF MAKING SECTIONS OF CORALS.[1]

(1) A small piece of the object is stained in any coloring fluid.

(2) Washed and treated with alcohol.

(3) Placed in a thin solution of gum copal in chloroform. (The copal is ground in a mortar with fine sand; the chloroform is poured over it, and the solution is then filtered.)

(4) The solution with the object is heated gently, in order to evaporate the chloroform.

(5) When the evaporation has proceeded so far that the solution can be drawn out in threads which become brittle on cooling, the object is taken out, and dried slowly by artificial heat until it is stony hard.

(6) Sawed with a fine saw into thin sections.

(7) Sections ground smooth and flat on one side with an ordinary whetstone.

(8) The section is then fastened to an object-slide by means of Canada balsam or the copal solution (must be warmed gently). When firm, it is

(9) Ground down, first with grindstone, then with whetstone, until the required thinness is attained.

(10) Washed and mounted in balsam.

The object can be decalcified by first freeing it from the copal or balsam with chloroform, then fixing it, by means of thick Canada balsam, on a slide, and then carefully decalcify, wash, stain, and mount as before.

This method might be employed in other cases where it is desired to preserve the relations between hard and soft parts; e. g., sections of star-fishes, bones, crinoids, etc.

SEMPER'S METHOD OF MAKING DRIED PREPARATIONS.

Semper's method, published in the Sitzungsber. d. phys.-med. Ges., Würzburg, 1880, and in the Zoolog. Jahresbericht for 1880, has been redescribed in detail by Dr. Sharp.[2]

---

[1] Zool. Anzeiger, No. II., 1878. Cf; Q. J. M. S. (N. S.), XX., p. 245, 1880.

[2] Proc. Acad. Nat. Sci. Philad., 1884, pp. 24–27.

1. Place the object in a weak solution of chromic acid (¼–2 %) 6–24 hours, according to its size and nature. For small animals, such as annelids, gastropods, frogs, mice, etc., 6–8 hours are sufficient.

2. Transfer to a large quantity of clean water, which must be often renewed until the acid has been so far withdrawn that the water remains uncolored by it. This part of the process may be much shortened by allowing a current of water to flow through the vessel. The usual time is from 10 to 20 hours.

3. Treat with 30–40 % alcohol 10–24 hours; with 60–70 % alcohol 2 or 3 days (with larger objects a week); with 90 % alcohol 2 or 3 days, or more; and finally with absolute alcohol.

The treatment with absolute alcohol is the most critical part of the whole process. *Absolutely* every particle of the water must be removed; for any tissue in which it remains will become spotted, and eventually spoil. Dr. Sharp always takes the precaution of changing the absolute alcohol once or twice, and leaves the object in it 3–7 days.

4. Transfer to turpentine, and leave it until it becomes thoroughly saturated (2–3 days). With large objects it is best to change the turpentine once.

5. Place the preparation in the air, in order to evaporate the turpentine, protecting it carefully from dust.

The preparation soon becomes white, resembling the whitest kid. It is light, stiff, and, on account of the resin contained, perfectly insect-proof.

If hollow organs (stomach, bladders, lungs, etc.) are prepared, they may be inflated with air after they have remained a short time in turpentine; by so doing, much space and consequently much alcohol are saved.

Prof. Semper keeps his preparations in dust-proof glass boxes, in which they can be seen from both sides.

To the five steps of the process, a sixth, discovered by Semper a few years ago, is given by Dr. Sharp. It consists in placing the prepared object in a solution of glycerine and sugar, which brings back almost entirely the original color in many cases.

O. P. HAY'S METHOD OF MAKING DRY PREPARATIONS.

While it may be true that in many cases the preparations made according to Semper's method have an appearance similar to a gypsum model, they quite often present a dingy, weather-beaten aspect that is by no means agreeable. The thin membranes and the connective tissues of dissections are left in a loose, woolly condition that grows worse by handling.

The microscopist completes his work by mounting his preparations in a solution of balsam. In like manner, Semper's method may be completed by saturating the preparation with some solid that will fill up the pores, bind the parts together, and restore the natural appearance. The solid which I have employed for this purpose is a mixture of Canada balsam, paraffine, and vaseline; but it is probable that a soft paraffine will, in most cases, do quite well. It is necessary that the mixture shall melt at about 46° C. (115° F.) While yet saturated with the turpentine, it is to be immersed in the mixture, heated a little above the melting point, and kept there until all the turpentine has been replaced. In many, if not in most, cases, however, the turpentine may be allowed to evaporate before the preparation is put into the melted paraffine mass. The latter then quickly penetrates the tissues, and the work is simplified. The preparation is then to be kept in an oven warm enough for the excess of paraffine to melt and drain off. It may then be wrapped in cloths, or in bibulous paper, until the whole of the paraffine mixture adhering to the outside has been dried off.

The advantages to be derived from pushing the process to this stage are the attainment of a greater degree of firmness and strength in the specimen, the obviation of the bleached appearance assumed on the escape of the turpentine, and the restoration of the natural colors. Probably any colors will reappear that will endure immersion in alcohol. In the case of anatomical preparations made in the way described, injected vessels show to advantage. I have also prepared specimens of lizards, small turtles, fishes, mussels, and earthworms; and whenever the tissues have been thoroughly saturated with the wax mass, the results have been satisfactory.

ON THE USE OF VASELINE TO PREVENT THE LOSS OF ALCOHOL FROM SPECIMEN JARS.[1]

The petroleum preparation known as vaseline is known to be practically unaffected by ordinary temperatures and by most substances. In the Journal of the Chemical Society, July, 1882, p. 786, it is said to be "sparingly soluble in cold, strong alcohol, and completely in hot, but separates out on cooling." After trying various substances — wax, paraffine, oil, and glycerine — with but partial success, the use of vaseline was suggested by the two authors independently and nearly at the same time. The experiments tried this spring indicate that, during three months, at ordinary spring and summer temperatures, there is no appreciable loss of 95 % alcohol from glass vials or jars, whether upright, or inverted, or on the side, provided corks are anointed on the bottom as well as on the side; provided ground-glass stoppers are anointed and firmly inserted; and provided the rubber rings of fruit jars and the specimen jars made by Whitall, Tatum & Co. are anointed on both sides, and the covers well screwed down.

We have also used the vaseline for preventing the loss of other liquids, excepting chloroform and spirits of turpentine; as a lubricator of drawers, and to prevent the sticking of the covers or stoppers of cement vials; and for the prevention of rust upon steel instruments.

---

[1] Wilder and Gage. Proc. A. A. A. S., XXXII, p. 318.

FORMULÆ FOR REAGENTS.

PRESERVATIVE FLUIDS, ETC.

ACID ALCOHOL (*Mayer*).

Alcohol (70 % or 90 %) . . . . 95 parts
Hydrochloric acid . . . . 3 "

For large objects.

BOILING ABSOLUTE ALCOHOL (*Mayer*).

For killing arthropods instantly, where other agents fail. —

GLYCERINE AND ALCOHOL (*Salvatore Lobianco*).

Glycerine . . . . 20 parts
Alcohol (70 %) . . . . 40 "
Sea water . . . . 40 "

KLEINENBERG'S PICRO-SULPHURIC ACID.

Picric acid (saturated solution in distilled water) . 100 parts
Sulphuric acid (concentrated) . . . . 2 "

Filter the mixture, and dilute it with *three* times its bulk of water; finally, add as much creosote as will mix. Use undiluted for Arthropoda (Mayer). Blanc dilutes with six times its bulk of water, and then adds a little 1 % acetic acid (2–3 drops for 15^{cc} of the solution) for Infusoria.

MAYER'S PICRO-NITRIC ACID.

Nitric acid (25 % $N^2 O^2$) . . . . 5 parts
Water . . . . 95 "
Picric acid, as much as will dissolve.

To be used instead of picro-sulphuric acid for objects possessing calcareous parts.

PICRO-CHROMIC ACID (*Fol*).

Picric acid (saturated aqueous solution) . . . . 10 parts
Chromic acid (1 %) . . . . 25 "
Water . . . . 65 "

Penetrates well only the smallest pieces of tissue.

PERENYI'S FLUID.

Nitric acid (10 %)	4 parts
Alcohol (90 %)	3 "
Chromic acid (½ %)	3 "

An excellent fluid for eggs that are brittle when hardened by other reagents.

MERKEL'S CHROM-PLATINUM SOLUTION.

Platinum chloride dissolved in water	1:400
Chromic acid	1:400

For the retina (Merkel); annelids (Eisig).

MODIFIED SOLUTION (FOR FISH OVA).

Platinum chloride	1:400
Chromic acid	1:100

Mixed in equal parts.

WITH ACETIC ACID (FOR PROTOZOA — *Brass*).

Platinum chloride	1 part
Chromic acid	1 "
Acetic acid	1 "
Water	400–1000 "

PERCHLORIDE OF IRON (*Fol*).

Perchloride of iron (alcoholic solution)	1 part
Alcohol (70 %)	5–10 "

Recommended for cilia, pseudopodia, small pelagic animals, etc.

CORROSIVE SUBLIMATE (Mercuric chloride).

Saturated aqueous solution (prepared by heating the water to boiling).

Employed where it is desirable to kill instantly in an expanded life-like condition; also much used for embryological purposes.

Goadby's Conserving Liquor:—

Aq. dest.	1,500	grms.
Salt	120	"
Alum	60	"
Sublimate	.25	"

Preserves marine animals well for a considerable time. Unsatisfactory for histological preservation.

Pacini's Fluids:—

A. Aq. dest.	200 parts
Salt	2 "
Sublimate	1 "

Serves for instantaneous killing.

B. Aq. dest.	300 parts
{ Salt	1 "
or	
Acetic ac.	2 "
Sublimate	1 "
Glycerine (25° Beaumé)	15 "

For nuclei, and dissolving red blood corpuscles (leaving the white).

IODINE SOLUTION (*Fol*).

Iodide of potassium (saturated aqueous solution).
Iodine (as much as will dissolve).

Water	100–500 parts

For *temporary* preparations.

Rabl's Fluids (*for nuclei*)[1]:—

A. Chromic ac. ($\frac{1}{2}$%)	200 g.
Formic ac.	4–5 drops

Causes chromatine filaments to swell somewhat.

B. Platinum chloride ($\frac{1}{2}$%).
Shrinks the filaments slightly.

Both fluids may be followed, after treatment with alcohol, by Grenacher's hæmatoxylin (dilute), safranin (after Pfitzner), or by hæmatoxylin and safranin.

FOL'S MODIFICATION[2] OF FLEMMING'S FLUID A.

Osmic ac. 1%	2 parts
Chromic ac. 1%	25 "
Acetic ac. 2%	5 "
Water	68 "

Erlicki's Fluid:—

Potassium bichromate	2$\frac{1}{4}$ g.
Cupric sulphate	$\frac{1}{4}$ g.
Aq. dest.	100 g.

---

[1] Morph. Jahrb., X., 2, p. 215, 1884.
[2] Lehrbuch d. vergl. Mik. Anat., p. 100, 1884.

Müller's "Eye-Fluid":—
    Aq. dest. . . . . 100 parts
    Bichromate of potassium . . . . 2 "
    Sulphate of sodium . . . . 1 "

Flemming's Fluids (*for nuclei*):—
    A.[1] Aq. dest. . . . . 100 parts
    Chrom ac. . . . . $\frac{1}{4}$ "
    Osmic ac. . . . . $\frac{1}{10}$ "
    Glacial acetic ac. . . . . $\frac{1}{10}$ "
For nuclear filaments omit the osmic acid.
    B.[2] Chrom. ac. (1%) . . . . 15 parts
    Osmic ac. (2%) . . . . 4 "
    Glacial acetic ac. . . . . 1 "
Stain with safranin or gentian violet.
    C.[3] Nitric acid (40–50 to aq. dest. 60–50).
Stain with Schneider's acetic acid carmine.

Strasburger's Fluid (*for nuclei*)[4]:—
    A. Acetic acid (1%) and a little methyl green.
Preparations are not permanent.
    B. Nitric acid (50%) followed by dye mixed in dilute glycerine.

Fr. Meyer's Preservative Fluid ("*Salicyl-Holzessig*"):[5]
    Pyroligneous acid (acetum lignorum) . . . . 100 parts
    Salicylic acid . . . . 1 "

This mixture is diluted with glycerine and water in three ways:—

1. *For Larvæ, Nematodes, Hydra, etc.:*
    Glycerine . . . . 10 parts
    Water . . . . 20 "
    Salicyl-pyrolig. ac. . . . . 3 "

2. *For Infusoria:*
    Glycerine . . . . 10 parts
    Water . . . . 40 "
    Salicyl-pyrolig. ac. . . . . 5 "

3. *For Algæ:*
    Glycerine . . . . 1 part
    Water . . . . 20 "
    Salicyl-pyrolig. ac. . . . . 1 "

---

[1] Zelsubst. Kern u. Zelltheilung, p. 382, 1882.
[2] Zeitschr. f. wiss. Mikroskopie, I., 3, p. 349, 1884.
[3] Arch. f. mik. Anat., XX., p. 1, 1881.
[4] Arch. f. mik. Anat., XXI., p. 476.
[5] Arch. f. mik. Anat. XIII., 1876, p. 868–869.

APPENDIX.

MACERATING AGENTS.

1. Alcohol . . . . 30% (or less)
2. Acetic acid . . . . 1% "
3. Osmic acid . . . . $\frac{1}{10}$% "
4. Chromic acid 05%–%.1

5. $\begin{cases} \text{Acetic acid} & \text{. . . . . 1 part} \\ \text{Osmic acid} & \text{. . . . . } \frac{1}{2} \text{ "} \\ \text{Sea water} & \text{. . . . . 1000 "} \end{cases}$

6. $\begin{cases} \text{Tincture of iodine} & \text{. . . . . 1 g.} \\ \text{Amniotic fluid} & \text{. . . . . 100 g.} \\ \text{Carbolic acid} & \text{. . . . . 1–2 drops} \end{cases}$

Fresh serum may be iodized by the addition of a few drops of a strongly iodized serum kept in stock.

7. $\begin{cases} \text{Eau de Javelle (KClO)} & \text{. . . . . . . 8 drops} \\ \text{Water} & \text{. . . . . 100 cc.} \end{cases}$

8. $\begin{cases} \text{Caustic potash} & \text{. . . . . 30–32 parts} \\ \text{Water} & \text{. . . . . 70–68 "} \end{cases}$

DECALCIFYING MIXTURE.

Chromic acid . . . . 70 parts
Nitric acid . . . . 3 "
Water . . . . 200 "

DESILICIFICATION (*Mayer*).

Fluorhydric acid added in drops to the alcohol containing the object.

STAINING MEDIA.

CARMINE.

Beale's Carmine.[1]

Carmine . . . . .6 grns.
Strong liquor ammoniæ . . . . 2 cc.
Glycerine . . . . 60 cc.
Aq. dest. . . . . 60 cc.
Alcohol . . . . 15 cc.

Having placed the carmine and ammonia together in a test-tube, heat over a spirit lamp, and the carmine will soon dissolve. The solution should be boiled for a few seconds, and

---

[1] How to Work with the Microscope, p. 109.

then allowed to cool. After the lapse of an hour, during which much of the excess of ammonia has escaped, the glycerine, water, and alcohol may be added and the whole filtered, or allowed to stand for some time, and the perfectly clear supernatant fluid poured off and kept for use.

ALCOHOLIC CARMINIC ACID (*Dimmock*).

Carminic acid . . . . 1 g.
Alcohol (80%) . . . . 400 g.

ALCOHOLIC AMMONIA CARMINE (*Dimmock*).

Alcoholic carminic acid.
Ammonia (added by drops until color changes to purple red).

ALCOHOLIC PICRO-CARMINE (*Dimmock*).

Alcoholic carminic acid.
Picric acid (dilute alcoholic solution) in very small quantity.

ALCOHOL CARMINE (*Mayer*).

Carmine (pulverized) . . . . 4 g.
Alcohol (80%) . . . . 100 cc.
Hydrochloric acid . . . . 30 drops

Dissolved by boiling 30 minutes; filtered warm and neutralized with ammonia.

PICRO-CARMINE (*Mayer*).

Carmine (powdered) . . . . 2 g.
Water . . . . 25 cc.
Ammonia (enough to dissolve carmine).
Picric acid (saturated aqueous solution — 4 times volume of the carmine solution).

Hoyer's Picro-Carmine:—

1. Dissolve 1 g. carmine in 6 ccm. ammonia, diluted with an equal volume of water.
2. Dissolve ½ g. picric acid in 50 ccm. of warm water.
3. Mix the two solutions and dilute so as to make 100 ccm.
4. Then add 1 g. chloral hydrate for preservation.

CARMINE ACETATE (*Böhm*).

Carmine (pulverized) . . . . 4 g.
Water . . . . 200 cc.
Ammonia (added by drops till the solution is deep red, and the carmine is dissolved).
Acetic acid (added until the solution turns to brick red).
Filter until no trace of a precipitate remains.

APPENDIX.

ACETIC ACID CARMINE (*Schneider*).

Acetic acid (45%) . . . . 1 part
Carmine (as much as will dissolve).
Water . . . . 99 "

GRENACHER'S CARMINE SOLUTIONS.

Alum Carmine: —
Alum (1–5% aqueous solution) . . . . 100 cc.
Carmine . . . . 1 g.

Boil 10–20 minutes, and filter when cool.

Acid Borax Carmine: —
Borax (1–2% aqueous sol.) . . . . 100 cc.
Carmine . . . . 1½ g.

Heat till carmine dissolves.

Acetic acid (added by drops while shaking, until the color is about the same as Beale's carmine).

Filter after standing 24 hours.

Borax Carmine: —
Borax (4% aqueous sol.) . . . . 100 cc.
Carmine . . . . 1 g.

Heat till carmine dissolves.

Alcohol (70% — equal to the vol. of carmine sol.)

Filtered after 24 hours.

Alcohol Carmine: —
Carmine . . . . 4 g.
Alcohol (60–80%, with 6–8 drops HCl) . . . . 100 cc.

Dissolve by boiling, then filter.

HÆMATOXYLIN.

GRENACHER'S HÆMATOXYLIN.

Hæmatoxylin (saturated alcoholic solution) . . . 4 cc.
Ammonia-alum (concentrated aqueous solution) . 150 cc.

Let stand 8 days in the light, then filter and add

Glycerine . . . . 25 cc.
Methyl alcohol . . . . 25 cc.

This dye works best after standing until a sediment forms.

WEIGERT'S HÆMATOXYLIN (*for nerves*).[1]

Hæmatoxylin	1 part
Alcohol	10 "
Water	90 "

Boiled and allowed to stand a few days.

WEIGERT'S DECOLORING FLUID.[2]

Borax	2 parts
Ferricyanide of potassium (red prussiate of potash)	2½ "
Water	100 "

Differentiates white and gray substance.

HEIDENHAIN'S HÆMATOXYLIN.[3]

Hæmatoxylin	½–1 g.
Water	100 cc.

Preceded by alcohol and followed by bichromate of potassium (½–1%). Stains nuclei black.

KLEINENBERG'S HÆMATOXYLIN.

Chloride of calcium (saturated sol. in 70% alcohol)	1 part
Alum (a little added, then filter)	1 "
Alcohol (70%)	6–8 "
Hæmatoxylin (saturated sol. in absol. alcoh., — enough to give the desired depth of color).	

MAYER'S COCHINEAL TINCTURE.

Cochineal (pulverized)	1 g.
Alcohol (70%)	8–10 cc.

Filter after standing several days.

CZOKOR'S ALUM-COCHINEAL.[4]

Cochineal	7 g.; pulverized with
Alum	7 g.; then add
Water	700 cc.; and boil down to 400 cc.

Add a trace of carbolic acid and filter. This stain may be used after any method of hardening; it differentiates well, giving the nuclei a hæmatoxylin shade, while staining the other parts of the cell red.

---

[1] Fortschritte d. Med., II., No. 6, p. 190, 1884.
[2] l. c.
[3] Arch. f. mik. Anat., XXIV., H. 3, p. 468, 1884.
[4] Arch. f. mikr. Anat., XVIII., p. 413.

APPENDIX.

ANILIN DYES.

Bismark Brown:—

Dissolved in boiling water, weak alcohol, 70% alcohol (**Mayer**), or dilute acetic acid (**Flemming**).

Stains well chromic acid or alcoholic preparations.

Eosin:—

Eosin . . . . 1 g.
Alcohol (95%) . . . . 100 cc.

For use, dilute with 20 parts alcohol (**Minot**).

Pfitzner's Safranin:—

Safranin . . . . 1 grm. dissolved in
Absolute alcohol . . . . 100 ccm.

After standing one or two days, filter, and dilute with

Distilled water . . . . 200 ccm.

PICRO-NIGROSIN (*Pfitzner*).

Picric acid (saturated aqueous sol.).
Nigrosin (small quantity).

Hardens and stains at same time.

PICRIC-HOFFMANN'S-BLUE.[1]

Dissolve Hoffmann's-blue in a saturated solution of picric acid in 50% alcohol, until the fluid assumes a dark greenish-blue tint.

DAHLIA (*Ehrlich*).

Alcohol abs. . . . . 50 cc.
Aq. dest. . . . . 100 cc.
Acid. acet. glac. . . . . 12½ cc.
Dahlia (to near saturation).

BLUE-BLACK (*Sankey*).

Blue black. . . . . ½ g.
Aq. dest. . . . . 1–2 cc.
Alcohol . . . . 99 cc.

SAFRANIN AND METHYL GREEN (*Ryder*).

Safranin (saturated alcoholic sol.) . . . . 1 part
Methyl green (saturated alcoholic sol.) . . . . . 1 "
Water . . . . 16 "

---

[1] Arbeit. Bot. Inst. Würzburg, III., pp. 53–60, 1884.

FIXATIVES.

ALCOHOL SHELLAC (*Giesbrecht*).

Bleached shellac	10 g.
Absolute alcohol	100 cc.

Leave 24 hours, then filter.

CARBOLIC ACID SHELLAC (*Mayer*).

Bleached shellac	50 g.
Absolute alcohol	250 cc.

Filter, and then evaporate the alcohol. Dissolve the shellac residue in

Carbolic acid	3–5 parts

COLLODION (*Schällibaum*).

Collodion	1 part
Clove oil	3–4 "

ALBUMEN (*Mayer*).

White of egg (filtered)	1 part
Glycerine	1 "

MOUNTING MEDIA.

GLYCERINE JELLY (*after Fol*).

Strong solution: —

Gelatine	30 parts
Water	70 "
Glycerine	100 "
Camphor (alcoholic sol.)	5 "

Weak solution: —

Gelatine	20 parts
Water	150 "
Glycerine	100 "
Camphor	15 "

CANADA BALSAM.

Prepared by heating on a bath until it becomes so hard on cooling that it cannot be indented by the finger-nail, and breaks like glass; then dissolved in an equal volume of *pure* benzole (beware of benzole that contains water).

DAMAR LAC.

Prepared by dissolving the gum in pure turpentine, or in a mixture of turpentine and benzole in equal parts.

APPENDIX.

TABLE I.
Comparison of the Units of Measure.

	1 MM. =	1 PARIS LINE =	1 ENGLISH LINE =	1 RHEIN LINE =	1 VIENNA LINE =
Millimeter	1.0000	2.2558	2.1166	2.1802	2.1952
Paris Line	0.4433	1.0000	0.9384	0.9964	0.9732
English Line . . .	0.4724	1.0659	1.0000	1.0299	1.0371
Rhein Line	0.4587	1.0347	0.9710	1.0000	1.0070
Vienna Line	0.4555	1.0275	0.9642	0.9930	1.0000

TABLE II.
Reduction of the Units of Measure to Micromillimeters.

MICROMILLIMETER.	PARIS LINE.	ENGLISH LINE.	RHEIN LINE.	VIENNA LINE.
$1\,\mu$ (0.001 mm.) = . . .	0.000443	0.000472	0.000459	0.000455
$2\,\mu$ = 0.002 mm. = . .	0.000887	0.000945	0.000917	0.000911
$3\,\mu$ = 0.003 mm. = . .	0.001330	0.001417	0.001376	0.001366
$4\,\mu$ = 0.004 mm. = . .	0.001773	0.001890	0.001835	0.001822
$5\,\mu$ = 0.005 mm. = . .	0.002216	0.002362	0.002293	0.002277
$6\,\mu$ = 0.006 mm. = . .	0.002660	0.002834	0.002752	0.002733
$7\,\mu$ = 0.007 mm. = . .	0.003103	0.003307	0.003211	0.003188
$8\,\mu$ = 0.008 mm. = . .	0.003546	0.003779	0.003670	0.003644
$9\,\mu$ = 0.009 mm. = . .	0.003990	0.004252	0.004128	0.004099
$10\,\mu$ = 0.010 mm. = . .	0.004433	0.004724	0.004587	0.004555
$20\,\mu$ = 0.020 mm. = . .	0.008866	0.009448	0.009174	0.009110
$50\,\mu$ = 0.050 mm. = . .	0.022165	0.023620	0.022935	0.022775
$100\,\mu$ = 0.100 mm. = . .	0.044330	0.047240	0.045870	0.045550

INDEX.

A.

Acetic acid, 17.
    for tracing nerves, 18.
    after staining, 182.
Acid, acetic, use of, 17.
    for decoloring, 46.
    alcoholic carminic, 37.
    carminic, 34-38.
    fluorhydric, 32.
    hydrochloric, with alcohol,
        for marine animals, 12.
        for bleaching, 16.
        for decoloring, 39, 46.
    nitric, 19.
    osmic, 15, 164.
    oxalic, 45.
        and alcohol, 25.
    picric, 15.
    picro-chromic, 22.
    picro-nitric, 21, 22.
Actiniæ, 28, 29, 203, 204.
Albumen, 122, 123.
    advantage of, **in staining with aniline
        dyes**, 123.
Alcohol, for killing, 5.
    for hardening, 6.
    the action of, 7.
    danger of maceration, 7.
    disadvantages attending its use, 12.
    Eisig's method of killing with, 12.
    acid alcohol for hardening, 13.
    boiling alcohol for killing, 13.
    with glycerine for killing, 13.
    method of making absolute, 14.
    iodine dissolved in, 15, 207.
    absolute, for hen's egg, 165.
Alcyonaria, 203.
Alosa, 177.
Amblystoma, 155, 156.
Amiurus, 179.
Amphibia, 155-162.
Amphictenidæ, 208.
Anæsthetic agents, 231.
Aniline dyes, 47-54.
    double staining with, 51.
    solubilities of, 54.
    not used with celloidin sections, 108.
    differential staining with, 123.
Annelids, 19, 198, 208.
Antiseptics,
    carbolic acid, 122, 211.
    chloral, 41, 43, 223.
    thymol, 43.

Apeltes, 178.
Aphides, 143, 144.
Appendicularia, 208.
Arthropods,
    eggs of, killed in hot water, 11.
    killed in boiling absolute alcohol, 13.
    hairs, etc., stained with osmic acid, 16.
    treatment with picro-sulphuric acid,
        20.
Artotype process, 196-199.
Asphalt-celloidin, 225.
Asterias, 147-149.

B.

Balanoglossus, 149, 150.
Barnacle, 174.
Batrachians, 196.
Beale's carmine, 38-39.
    macerating action, 29.
Belone, 177.
Bergamot oil, 109, 114, 127.
Bismarck Brown, 48, 51, 52.
Bleaching,
    after osmic acid, 16.
    natural pigment, 17.
    after gold chloride, 56-57.
    after silver nitrate, 58.
Bleu de Lyon, 48-49.
Blue black, 50.
Brachiopoda, 126, 209.
Bryozoa, 232.

C.

Calberla's fluid, 163.
Caoutchouc, 120.
Capitellidæ, 23.
Caprellidæ, 53.
Carbolic acid, 122, 211.
Carbonic acid as an anæsthetic, 231.
Carminate, alcoholic ammonic, 37.
Carmine, alcohol (Mayer), 39, 123.
    alum (Grenacher), 39.
    acid-borax (Grenacher), 40.
    borax (Grenacher), 40.
    acetic acid (Schneider), 43.
    emulsion, 221.
    solutions, 34-44.
Carminic acid, 34-38.
Cat-fish, 179.
Celloidin, 107-114, 117, 120.
    advantages of, 112-113, 127.
    injections, 225.

INDEX.

Central nervous system,
  Weigert's method, 191–192.
  method of counting fibres and ganglia, 192–194
    Betz's method, 194.
    Osborn's method, 194–196.
    photographing sections, 196–198.
  of annelids, 198, 199.
Chirostoma, 177.
Chloral, 41, 43, 223.
Chloride of calcium, 44.
Chloroform, 6, 95, 96, 114.
Chlorophyll, 212.
Chlor-zinc-iod, 212.
Chondrosia, 208.
Chromic acid, for killing, 6.
  principal drawback and best means of meeting it, 6, 9.
  use of, 18.
  for nuclei, 19.
  combined with other acids, 19.
  precipitates avoided by keeping dark, 19.
  for decalcification, 31.
  for frog's egg, 157, 158.
  hot, for hen's egg, 166.
Clarifying media, 94.
  penetration hastened by warming, 94.
  mode of avoiding shrinkage, 94.
  chloroform preferred to creosote or oil of cloves, 95.
  Andres' mixture, 95.
  for celloidin sections, 109.
  clove oil followed by turpentine, 94.
  benzole, 142.
  creosote, xylol, 192.
Clepsine, 136–138.
Clove oil, 94, 109, 113, 182.
  followed by turpentine, 94.
Cocaine, chlorhydrate of, 231.
Cochineal, 12, 35, 41.
  (Mayer), 45–47.
Cod, 176, 177.
Cœlenterates, 29, 231.
Collodion,
  method of fixing sections, 117.
  as a fixative, 121, 122.
  uses of, 126–128, 170–172.
Comatulæ, 139.
Corals, sections of, 233.
Corpus vitreum, 209.
Corrosive sublimate, 5, 7, 15, 18, 28, 37, 134.
  use of, 26.
  removal of, facilitated by camphor, 204.
  combined with Perenyi's fluid, 207.
Crab, 175.
Crangon, 176.
Crayfish, 141.
Creosote in picro-sulphuric acid, 20, 21.
Ctenophora, 209.
Cyanide of potassium, 56, 58.
Cyanin, 53.

D.

Dahlia, 49–50.
Decalcification, 31.
Decoloring,
  with hydrochloric acid, 39, 45, 46.
  with ammonia, 39.
  with oxalic acid, 45.
  with 70 per cent alcohol, 46, 181, 182.
  with acetic acid, 46, 49.

Decoloring—*continued.*
  with alcohol, 48.
  with HCl and alcohol, 142.
Desilicification, 32.
Dicyemidæ, 16.
Diemictylus, 156.
Doliolum, 150, 151.
Drawing apparatus, 129–134.
Dried preparations,
  Semper's method, 233.
  Hay's method, 235, 236.
Dripping apparatus, 115.

E.

Echinoderms, 21, 201, 231.
Echinorhynchi, 205.
Egg-emulsion, 106, 107, 109, 110.
  bath for hardening, 105.
Elder pith, 106, 109.
Embryological methods, 129–173.
Emulsion, carmine, 221.
  blue, 222.
  black, 222.
Eosin, 48, 51.
  Ogatha's, 210.
Erlicki's fluid, 24, 25, 187, 191.

F.

Ferricyanide of potassium, 56.
Fibrospongiæ, 32.
Filtering bottle, 55.
Fishes, fatty pigments, 16.
  eggs of, hardened in Merkel's fluid, 19.
  nitric acid for hardening the brain, 19, 172.
  Perenyi's fluid, 23.
  osmic acid and Müller's fluid for the eggs, 24.
  pelagic fish eggs, 151–154.
  trout's egg. 162.
  eggs, 176–179, 208.
  corpus vitreum, 209.
Fixatives, for serial sections, 116–123.
  alcoholic shellac, 116, 117.
  carbolic acid shellac, 117, 118.
  Gaule's method, 119.
  gutta-percha, 119, 120.
  caoutchouc, 120, 121.
  collodion, 121, 122.
  albumen, 122, 123.
Fluorhydric acid, 32.
Formulæ for reagents, 237–246.
Fuchsin, 51.

G.

Gambusia, 178.
Gammarus, 176.
Ganglion cells, method of counting, 194.
Glycerine and alcohol for killing, 13, 204.
  and gelatine for imbedding, 106.
Glycerine jelly for mounting, 125.
Gold chloride, 56, 187–190.
  Flechsig's method, 187.
  Freud's method, 187–188.
  Gaule's method, 188.
  Ranvier's methods, 188–189.
  Ciaccio's method, 189.
  Pfitzner's method, 190.
  after chromic acid, 188.
Gold-fish, 179.
Gravitation, influence of, on cleavage, 159–162.

Gum arabic, for imbedding, 28, 106, 109.
Gutta-percha, 119, 120.

H.

Hæmatoxylin, 12.
  (Grenacher), 44, 184.
  (Kleinenberg), 44, 45.
  Heidenhain's method, 186.
  Weigert's method, 191.
Hardening, 6, 7, 11.
  with acid alcohol, 13.
  with boiling alcohol, 13.
Hatching jar, 177, 178.
Helix, 145.
Hen's egg, 19, 163-167.
Hippa, 174.
Histological methods, 200-212.
Hoffman's blue, picric, 212.
Hoffman's violet, 51.
Hydra, 186, 232.
Hydrochloric acid,
  for removing ferric salts, 25.
  for decalcifying, 31.
  for decoloring, 39.
Hydroids, 208.
Hyperia, 174.
Hyposulphite of sodium, 58.

I.

Imbedding, methods of, 93-115.
  ribbon method, 94.
  melting point of paraffine, 93, 94.
  preparation for, 94, 142.
  apparatus, 102-104, 109.
  boxes, 97, 108, 111, 112.
  orientation, 98, 99.
  prevention of bubbles, 99, 100, 113.
  Schulgin's mixture for, 100.
  Caldwell's method of, 100, 101.
  Ryder's method of, 101.
  in gum arabic, 106, 109.
  in glycerine and gelatine, 106.
  in egg emulsion, 106, 107, 109, 110.
  in elder pith, 106, 109.
  in liver, 109.
  in celloidin, 107-109, 111-115.
Infusoria, 19, 25, 53, 202.
Injection,
  methods of, 221-231.
  dry masses, 221-223.
  starch, 223.
  special mass, 224.
  dried preparations, 225.
  celloidin, 225.
  double injections, 225-231.
  apparatus for double injections, 229.
Iodine, use of, 14, 15, 30, 207.
Iodized serum, 30.

K.

Karyokinetic figures, 15, 19, 44, 50, 181-186.
  as a guide to intensity of growth, 183.
  chrom-formic acid contrasted with platinum chloride, 184.
Killing, general statements regarding, 4, 11.
  planarians, 5.
  annelids, 5.
  with hot water, 11, 157.
  with alcohol, 12.
  with boiling alcohol, 13.

Killing — *continued*.
  with glycerine and alcohol, 13.
  with perchloride of iron, 25.
  with corrosive sublimate, 26-27.
  with nigrosin and picric acid, 49.
Kleinenberg's picro-sulphuric acid, 5, 7, 18.
  not a hardening fluid, 11.
  composition of, 19-21.

L.

Leeches, 27
Lemon juice, 189.
Lepidosteus, 127.
Limax, 145-147.
Limulus, 144, 145.
Lineus, 175.
Lobster, 174.
Loligo, 176.
Loxosoma, 207, 208.
Lumbricus, 47.

M.

Maceration, 28-31.
  Andres' method with Actiniæ, 28, 205.
  the Hertwigs' mixture for cœlenterates, 29.
  with iodized serum, 30
  with eau de javelle, 30.
  with caustic potash, NaCl., 211.
  with alcohol, 33⅓ %, 211.
  with chloral hydrate, 211.
  with chromate of ammonia, 211.
  with chromic acid and chloral hydrate. 211.
Meckelia, 175.
Medusæ, 201.
Menidia, 178.
Merkel's chrom-platinum solution, 5, 7, 18, 19.
  arrests blackening by osmic acid, 16.
  composition and use, 13.
  modified form of, 153, 200.
  for nuclei, 184.
Methods, macroscopical and microscopical, 1.
Methyl green, 52.
  violet, 51.
  and gentian violet, 53.
Methylen blue, 51.
Meyer's fluid, 31.
Micrometer-screw, 64.
  registering, 66.
Microtome, the, in zoology, 2, 10.
  price of, 69.
  two types of, 60.
  Thoma's, 61-69.
  Schanze, 69, 70.
  Caldwell's, 70-74.
  care of, 76.
  the freezing,
  Rutherford's, 76-79.
  Jacob's, 79-80.
  ether-freezing apparatus, 80, 81.
  notes on, by Gage, 81, 82.
  gelatine jelly instead of gum, 83
Microtome knives, 83-86.
  method of sharpening, 85, 86.
Microtomist, 1.
Microtomy, 2.
Molgula, 176.
Mollusca, 36, 145-147.
Molybdic acid test, 212.

Monacanthus, 178.
Mounting media, 124, 125.
    balsam in xylol, 119.
    balsam in chloroform and clove oil, 122.
    balsam, solvents of, 119, 124.
        preparation of, 125.
    damar, 124, 125.
        preparation of, 125.
    glycerine jelly, 125.
Mucous membrane of stomach, 211.
Müller's fluid, 24, 191, 211.
Muscular fibres stained with palladium chloride, 59.
Museum methods, 231–236.
Myriapods, 179, 180.
Myrtillus, 50, 51.
Mysis, 17.
Myzostoma, 139, 140.

N.

Naphtha, 122.
Neophalax, 141–143.
Nerve-endings, 18, 19, 188, 189, 190.
Nerve-fibres, 58, 193.
Nerves of cornea, 59.
Nicotine, 204.
Nigrosin and picric acid, 49.
Nitrate of silver, 57, 58.
    with marine objects, 58, 207.
    Golgi's method for nerve-fibres, 190, 191.
    Hertwig's method for marine animals, 209.
Nitric acid, 19, 172.
    combined with chromic acid for decalcifying, 31.
    for nuclei, 182.
Nuclei, 181–186.
    stained with safranin, 49.
    " " dahlia, 49.
    differentially stained, 123.
    in general, 181, 182.
    ova of Echinoderms, 182.

O.

Object-holder and carrier, 67–69.
Oil for microtome, 64, 76.
Opaque-celloidin, 225.
Origanum oil, 113, 127.
Orientation,
    in imbedding, 98, 99.
    of frog's eggs, 161.
    of hen's egg, 163–165.
    in sections, 213.
Osmic acid, 15.
    as a staining medium for hairs, etc., 16.
    blackening process arrested, 16.
    bleaching after, 16, 17.
    for strengthening picro-chromic acid, 22.
    for hen's egg, 164.
    after silver nitrate, 208.
    followed by silver nitrate, 209.
Oxalic acid, 26.
Oyster, 52.

P.

Palinurus, 22.
Palladium chloride, 58, 59.

Paraffine,
    cooled in water to prevent air-bubbles, 71.
    for imbedding, 93–104.
    saturation with, 96.
    solvents of, 94–96, 122, 123, 142.
    and vaseline, 120.
    and ceresin, 100.
Pedicellina, 207.
Perca, 179.
Perchloride of iron, 25.
Perenyi's fluid, 19, 22, 154.
Phronima, 16, 174.
Phryganids, 141–143.
Picric acid, 15.
    for decalcification, 32.
    for making picro-carmine, 38.
    combined with nigrosin, 49.
Picro-carmine (Ranvier), 41, 42.
    (Pergens), 41.
    (Minot), 42.
    (Mayer), 42, 43.
Picro-chromic acid, 22.
Picro-nitric acid, 21, 22.
Planarians, 134, 135.
Planorbis, 145.
Polygordius, 198.
Polyxenes, 180.
Potash, caustic, 31.
Potassium bichromate, 23.
Prawn, 174.
Preservative fluids, 11.
Protodrilus, 198.
Protoplasm, continuity of, 212.
Protozoa, 200, 201.
Pseudo-chlorophyll bodies, 186.

R.

Rabbit, 167–170.
Radiolaria, 32.
Rana, 155.
Reagents,
    formulæ, 237–246.
Reconstruction, from sections—
    His and Born 2, 215–218.
    general remarks on, 213–215.
    illustration, 213, 215–218.
Reptiles, 198.
Reptilian ova, 162.
Rhizopods, 202.
Rose Bengale with Iodine Green, 48.
Rosein, 51.
Rossia, 176.

S.

Saccocirrus, 198.
Safranin (Pfitzner), 49.
    with methyl green, 52.
Sagitta, 140, 208.
Salmonidæ, 154–155.
Salt solution, 143, 207, 211.
Sapphirina, 16, 17.
Sectioning, order of steps preparatory to, 4.
    combination, suppression, and transposition of steps, 4.
Section-observation, 2, 10.
Sections, ribbon method, 2.
    cut under alcohol, 106. 113–115.
    dripping apparatus, 115.
    fixatives, 116–123.
    collodionized, 127, 166.

Sections—continued.
    prevention of rolling, 142, 143, 170–172.
    orientation, 213.
    of corals, 233.
Section-smoothers, 86–92.
    of Mayer, Andres, and Giesbrecht, 87, 88.
    Schulze's, 88, 89.
    Decker's, 90.
    Kingsley's, 91.
    of Gage and Smith, 91, 92.
Shad, 177.
Shellac, alcoholic, 116, 117.
    carbolic acid, 117, 118.
Slide devised by Rabl, 185.
Skeletons prepared in eau de javelle, 31.
Spinal cord,
    staining with rose Bengale, 48.
    "   "   palladium chloride, 59.
Spirorbis, 138, 139.
Sponges, 201, 208.
Spongilla, 31.
Staining, 8.
    alcoholic and aqueous solutions, 8, 33.
    qualifications of a dye for staining in toto, 8.
    with gallic acid, 25.
    methods of, 33–54.
    intra vitam, 52.
    Haidenhain's method, 211.
Stickleback, 178.
Stomach, mucous membrane of, 211.
Surface-observation, 2, 10.

T.

Teleostei, 151–153.

Temperature, methods of keeping uniformly cool, 136, 151.
Tergipes, 175.
Thymol, 43.
Tobacco smoke, 204.
Tomopteris, 208.

U.

Urodela, 194–196.

V.

Vaseline,
    to prevent loss of alcohol from specimen jars, 236.
Vesuvian-celloidin, 225.
Vesuvin, 51.

W.

Water, hot, for killing, 11.
Water-bath, 104, 105.
White perch, 179.
Wickersheimer's fluid, 225, 227, 232.

X.

Xylol, 119, 122, 192.

Y.

Yellow perch, 174.

Z.

Zoochlorella, 186.

AUTHORS.

ABBE, 200.
Agassiz, 139, 147-149, 151, 152
Altmann, 30.
Andres, 2, 28, 66, 87, 95, 98, 140, 203, 204.

BACHMANN, 125.
Bachrach & Brothers, 108.
Balfour, 163.
Bastian, 59.
Bateson, 149, 150.
Beard, 139.
Bemmelin, 209.
Beneden, Ed. van, 16, 167-170.
Berthold, 14.
Betz, 194.
Bianco, 204.
Birge, 24, 192-194.
Bizzozero, 53.
Blanc, 201.
Blanchard, 26.
Blochmann, 97, 108, 113, 127.
Boecker, 81.
Boehm, 43, 155, 163.
Böttcher, 47, 181.
Born, 2, 160, 161, 170, 216.
Brandt, 52, 53, 186.
Brass, 19, 200, 201.
Brooks, 176.
Browning, 223.
Bütschli, 95, 96.
Bunge, 109.

CALBERLA, 106, 107, 109, 158, 163, 195.
Caldwell, 2, 61, 70, 86, 100.
Carnoy, 42.
Certes, 53, 201.
Ciaccio, 189.
Cybulsky, 56, 57.

DECKER, 2, 90.
Dimmock, 34-38, 97, 98.
Dorp, van, 26.
Duval, 111, 126, 127, 163-166.

EISIG, 5, 12, 19, 23.
Emery, 16.
Engelmann, 19.
Entz, 201.
Erlicki, 24, 25.

FASCHING, 89.
Faxon, 174, 175.
Fewkes, 138, 139.
Flechsig, 56, 187.

Flemming, 18, 19, 47, 123, 181-183, 186.
Fol, 14, 18, 19, 20, 21, 22, 24, 25, 41, 59, 85, 99, 125, 145, 219, 231.
Fraipont, 198, 199.
Francotte, 91, 116.
Frenzel, 119, 120.
Freud, 25, 187.

GAGE, 81, 82, 91, 121, 127, 223, 224, 236.
Gannett, 115.
Gardiner, 212.
Gaule, 116, 119, 188, 192.
Gerlach, 34.
Giesbrecht, 2, 27, 66, 87, 94, 95, 96, 98, 116, 119, 127.
Girard, 147.
Golgi, 58, 190.
Goronowitsch, 172.
Gravis, 106.
Grenacher, 35, 39, 110.
Griesbach, 48.

HANAMAN, 55.
Harmer, 58, 207.
Harris, 51, 54.
Hay, 227-231, 235, 236.
Heidenhain, 24, 186, 211, 212.
Heitzmann, 57.
Henle, 59.
Henneguy, 24, 154, 155.
Hermann, 47, 181.
Hertwig, Oscar, 29, 107, 140, 157-159, 162, 209.
Hertwig, Richard, 29, 58.
His, 2, 19, 58, 129, 215.
Hoffman, 97, 102, 103, 137.
Hughes, 64.

IIJIMA, 134, 135.

JACOB, 79, 80.
Julin, 169.
Jung, 60, 69, 86, 113.

KAISER, 106.
Kingsley, 91, 117, 144, 145, 175, 176.
Körting, 69.
Koch, 233.
Kölliker, 18, 163.
Korschelt, 201.
Kossman, 65, 96.
Kleinenberg, 19-21, 44, 45, 47.
Krieger, 215.
Kronecker, 169.

Kupffer, 155, 162, 163.
Kutschin, 211.

LANDSBERG, 201.
Lang, 26, 27.
Lavdowsky, 50.
Legros, 57.
Lewes, 81.
Lewis, 64.
Liebermann, 36.
Lockwood, 144.
Loewit, 189.

MARK, 29, 116, 127, 128, 145-147, 170-172.
Mason, 2, 127.
Mason, J. J., 196.
Max Flesch, 24, 192.
Mayer, Paul, 2, 11, 12, 16, 17, 20-22, 27, 32, 33, 39, 40, 45, 46, 48, 53, 66, 87, 98, 104, 116-118, 122, 123, 127, 208.
McLachlan, 142.
Merkel, 23, 59, 111, 126.
Meyer, 208, 223.
Minot, 19, 24, 39, 42, 48, 76, 94, 113-115, 186, 192.
Moeller, 61.
Morse, 136.
Müller, 24.

NOLL, 30.

OGATHA, 210.
Osborn, 24, 194-196, 225-227.
Oschatz, 60.

PACKARD, 149.
Pansch, 223.
Patten, 86, 141-143.
Perenyi, 22, 23, 154.
Perls, 30.
Pfitzner, 19, 190.
Pflüger, 159, 160.

RABL, 19, 184, 185.
Rabl-Rückhard, 19, 172.
Ransom, 207.
Ranvier, 14, 30, 40, 57, 109, 183, 189.
Rathke, 141.
Rauber, 162.
Redding, 56.
Reichenbach, 141.
Reichert, 61.
Richard, 232.
Rindfleisch, 109.
Rivet, 60.
Rollett, 212.
Rosenberg, 215.
Rossa, 223.
Roux, 161.

Roy, 81.
Rue, De la, 35, 36.
Ruge, 107, 109.
Rutherford, 76, 82.
Ryder, 52, 101, 102, 176-180.

SÆFFTIGEN, 205, 206.
Sankey, 50.
Schäfer, 30, 32, 125.
Schällibaum, 2, 116, 121, 122, 127.
Schaller, 35, 36.
Schanze, 60, 61, 69.
Schiefferdecker, 107, 108, 111, 112, 126, **128**, 225.
Schulgin, 100.
Schulze, 2, 58, 59, 88.
Schweigger-Seidel, 35, 40.
Scott, 107.
Sedgwick, 94.
Seessel, 215.
Semper, 233.
Sharp, 14, 127, 233, **234.**
Smith, S. I., 175.
Smith, 91, **121.**
Sollas, 83, **208.**
Stimpson, **149.**
Stöhr, 217.
Strasburger, **19,** 183, **184.**
Swirski, 217.

THANHOFFER, 59.
Thoma, 61-65, 109-113, 127.
Threlfall, 120, 121.
Timm, 127.
Trinkler, 211, 212.

ULJANIN, 150, 151.
Uskow, 217.

VERRILL, 149, 175, 176.
Virchow, H., 19, 209.
Vosmaer, 208.

WATTS, 34, 35.
Weigert, 24, 25, **191, 192.**
Weismann, 95.
Whitman, C. O., **19,** 136-138.
Whitman, E. N., 174, **175.**
Wikszemski, 223.
Wilder, 224, 236.
Will, 127, 143.
Wilson, 136, 144, 203.
Witlaczil, 143.
Wittich and Benkendorf, 108.

ZADDACH, 141.
Zeiss, 60, 69.
Zimmerman, 83.

www.ingramcontent.com/pod-product-compliance
Lightning Source LLC
Chambersburg PA
CBHW032135230426
43672CB00011B/2344